Earthworms, Horses, and Living Things

Earthworms, Horses, and Living Things

William DuPuy's
Our Animal Friends and Foes

Edited and introduced by
Paul Rich

WESTPHALIA PRESS
An imprint of Policy Studies Organization

Earthworms, Horses, and Living Things:
William DuPuy's
Our Animal Friends and Foes

Westphalia Press
An imprint of Policy Studies Organization
dgutierrezs@ipsonet.org

For information:
Westphalia Press
1527 New Hampshire Ave., N.W.
Washington, D.C. 20036

ISBN-13: 978-0944285985
ISBN-10: 0944285988

Updated material and comments on this edition can be
found at the Policy Studies Organization website:
http://www.ipsonet.org/

This edition is dedicated to Joshua Gannon,
master of layout and typography

Also from Westphalia Press

THE GIFTED WILLIAM DUPUY
PREFACE TO THE NEW EDITION

───────────

WILLIAM Atherton DuPuy (1876-1941) deserves more than a dusty shelf in an antiquarian bookstore. Few American writers about nature had the facility for making the subject intelligent and interesting at the same time with the facility that he did.

He was fastidious about checking sources and getting expert opinions, but when he had a good story he knew how to use it. His classic account in these pages of the role of garden slugs in detecting mustard gas attacks during World War I is a case in point.

Despite all the years that have passed, this book remains topical. Recently, there has been much written about the urban invasion, or perhaps we should say reinvasion, by a variety of animals that have found our garbage cans to be an inviting

E

buffet. Opossums are under the back porch and bears meander down streets and break into cars. Meanwhile, bats are threatened by diseases introduced into their subterranean habitats by spelunkers and fresh water mussels are having a hard time with our toxic wastes.

Possibly an alternative title for this volume would be "Our Animal Friends and Victims". In any case, DuPuy's cast of animal characters are still very much part of our lives.

Paul Rich

THE OPOSSUM, A NIGHT MARAUDER

Romance of Science Series

OUR ANIMAL FRIENDS AND FOES

By
WILLIAM ATHERTON DuPUY

Introduction by
EDWARD WILLIAM NELSON
CHIEF, UNITED STATES BUREAU OF
BIOLOGICAL SURVEY

Illustrated by
EDWARD HERBERT MINER
ANIMAL PAINTER TO THE NATIONAL
GEOGRAPHIC SOCIETY

THE JOHN C. WINSTON COMPANY
CHICAGO   PHILADELPHIA   TORONTO
ATLANTA   SAN FRANCISCO   DALLAS

PREFACE

SURROUNDED as we are by animal life, it is none the less true that few of us ever stop to fit the beasts about us into their proper niches, to understand what they are, how they came about, what is the purpose they serve.

Few of us know, for instance, that the tabby cat is an Egyptian; that the frog invented arms and legs; that the whale was once a land animal walking on four feet; that the earthworm is probably more important to man than the horse; how the milk cow came about; that toads and bats are among man's best friends; and that the house rat has caused the death of more men than war.

We do not stop to learn much about the commonplace animals which are around us. This, in fact, has not been particularly easy.

Biology, the study of living things, has always been with us, to be sure, but in a form so weighty, surrounded by such technicalities of language, accompanied by such infinite detail, that few of us have had the hardihood to attempt its mastery.

Animals occupy a place of primary interest in the human consciousness. Few people will pass the nose of a friendly Dobbin, standing at the curb, without stopping to rub it. There are few to whom the goldfish in its bowl, the canary in its cage, have not a charm. Newspaper editors believe that stories of dogs are among the most appreciated items that they can print.

Animals have life, are given to action; and life and
action have first claim on human attention. When I
started to write a book on animals I bore this fact in
mind. I sought to gratify the natural curiosity which
man, standing at the top of the list of animals, has about
his humbler friends. I have tried to write a book which sets
forth primarily the throbbing interest of the animal world.

It has been a part of that interest that each animal
should be fitted into its proper niche, and that it should be
set forth in its relations to other animals and to the world at
large. My primary purpose has been to awaken an interest
in Biology, to give a bit of a bird's-eye view of that science,
and, at the same time, to furnish genuine entertainment.

I have, therefore, taken infinite pains that what I have
written should be in accord with scientific fact. I have
appealed for aid and guidance to that greatest agency in
the world for the study of animals, the Biological Survey
of the United States Government.

The Biological Survey studies animal life from the
practical standpoint of its bearing on the welfare of the
people. Its province is the conservation and upbuilding
of the supply of useful and otherwise valuable forms of
wild bird and animal life, including those classed as game
and as fur bearers, and the control of species such as
wolves and mountain lions, which harass the stockman
of the West, and of rodents, which through their vast
numbers, become pests injurious to agriculture, forestry,
and stock raising. It is the official Biological Bureau of
the government; but it studies Biology from a practical
standpoint, from the standpoint of its direct bearing on
the national welfare. It has studied animals, not merely

as scientific specimens, but in their relation to the world community of which they are a part.

The Biological Survey develops the sort of information that gives purpose to a knowledge of animal life.

Dr. Edward William Nelson has been the head of this organization for many years. He has been a naturalist in the public service for almost half a century. He began as a mere youth with four years of exploration in Alaska, then a virgin field. He served as naturalist on the U. S. Revenue Service Cutter, *Corwin*, in her expedition into the Arctic, away back in 1881. As though seeking the greatest possible climatic contrast, he later took part in an expedition into Death Valley, California. He devoted fourteen years to field research in Mexico, became the government's chief field naturalist, and finally assumed the highest post it can offer to one of his profession.

As I wanted to write a book of stern merit, it was to this Bureau and to this man that I appealed for supplementary aid. All the facilities of the Bureau were put at my disposal. The government is interested in education, in getting the true message to the multitude.

I was counseled as to the reliability of authors. Many books, be it said, are written on a basis of secondhand knowledge. The authors are not able to weigh the reliability of statements gained from reading.

To what I had known of animals I have added by very extensive reading. I have massed my facts, written my chapters. I have brought these chapters back to the Biological Survey for reading, criticism, and editing. They have been put under the microscope of scientific men who specialized on each animal, who knew each from years of study of it in the field. Sometimes I have sent these chapters outside the Bureau for criticism by some outstanding authority. The chapter on the earthworm, for instance, went to Dr. Frank Smith, of the University of Illinois, who is admittedly the best American authority on the subject; the chapter on the coral polyp went to Dr. Paul Bartsch, of the National Museum; that on rattle-snakes to Dr. Leonhard Stejneger, curator of reptiles for the same institution; the ones on fresh water mussels and fur seals to the Government's Bureau of Fisheries, and so on.

But all were read and edited by Dr. Nelson. He had hunted whales in the far north and monkeys in Guate-mala, and he gave me the benefit of his experience. He had caught rattlesnakes in Arizona for scientific use, and explained his personal method of making sure that they were dead before putting them in the game bag. He caught them close up to the head, he said, held them tight, and when they opened their mouths wide to bite, he thrust the blade of his penknife into the mouth and severed the spinal cord at the base of the skull. He brought personal experience in the field into the consider-ation of every chapter.

Under such guidance and with such assistance I have made a book in which I have confidence. There are none of the mistakes in it that are likely to creep into the writ-ings of theorists or into anything for which one man alone is responsible. The auspices under which the book has been created have been so favorable, and the aid lent so exceptional, that I feel I may, without immodesty, sing its praises.

WILLIAM ATHERTON DU PUY

CONTENTS

ILLUSTRATIONS

INTRODUCTION

DURING the gradual development of civilization, with the attendant domestication of animals and cultivation of crops to replace Nature's wild products as sources of food, clothing, and other necessary comforts and luxuries, most people have lost nearly all of their ability to see a large share of the wood folk that come within the range of their vision. Thus they know little of their doings. Even when they casually note one of the wild habitants, they frequently have no conception of its identity or the meaning of its activities. They need an interpreter to enable them to observe and understand what is quickly apparent to the small minority among mankind fortunate enough to retain from a more primitive existence much of their ancestral powers of seeing and understanding.

Apparent miracles are commonplaces in Nature. All about us marvelous happenings occur, and on every hand the most fascinating adventures take place among our big and little fellows of the wild. These are all open to the eyes of those having the ability to understand and appreciate what they see. In early days primitive man lived in caves and huts in the wilderness, and his very life was often dependent on his intimate knowledge of wild things and their ways. That knowledge, however, has been largely lost.

During many years of field experience as a naturalist and hunter of large and small game, I have noted with much interest the wide variation of men's powers of observation, both of sight and of understanding of what is visible in the manifestations of Nature with which they come in contact whenever they leave the densely occupied places. Some men have learned to see the life about them and some have remained blind to it. A new world has been opened to those who have learned how to look into the lives of birds or beasts and much pleasure has come to them throughout their lives.

Even among field naturalists comparatively few have the ability to be all-round observers and interpreters of Nature. Many so specialize their powers of observation that they become practically blind to the things outside their own limited field. A striking illustration of this was once demonstrated to me when with a scientific party in a picturesque mountain canyon of California. We camped at a beautiful spot on the bank of a clear, sparkling mountain brook, flowing through small grassy flats. Work in hand kept me in camp during the morning hours of the first day, but an eminent botanist of the party went up the canyon for some distance, collecting plants, and returned about noon with a portfolio loaded with interesting specimens. My particular interest was in securing a series of specimens of chipmunks from this vicinity for scientific study and I inquired of the botanist whether, during his morning walk, he had noted any of these animals. He considered the question thoughtfully for a moment, then replied that he had not. This was a disappointment to me, but in the afternoon when I went up the canyon in search of birds and mammals, I was highly pleased to find chipmunks in great abundance.

Later trips proved that at all hours of the day these grace-ful little animals were much in evidence. Despite this, the botanist had been so intent upon his own plant obser-vations that the chipmunks, in which he had no interest, had made not the slightest impression on his mind, al-though the image of many of them must have been received by his eyes.

This kind of mental, in place of ocular, blindness is really existent among a vast majority of mankind, and as a result they know only the more conspicuous and striking birds and animals which appear directly in their way. Here is where the all-round observing powers of natural-ists, either of the professional or the amateur class, come in to qualify them as the seers and interpreters of Nature.

Such successful writers on wild life as Mr. Du Puy are of the utmost importance in helping to build up an intel-ligent appreciation of Nature through their ability to bring to the public in a delightful way many interesting facts about our fellow habitants of the world. Mr. Du Puy is especially gifted in narrating the lives of many varied species of animals from the lower to the higher forms. His work is a direct help and renders his readers a real service by enlarging their powers of obser-vation.

The last twenty-five or thirty years in the United States has seen an enormous growth of interest in Nature and in the ways of wild things. A multitude of people are learning to see, more or less clearly, some of the wonders and beauties about them and are individually becoming more and more capable of interpreting what they see.

No one, whether a school child, or a person of many years, can enter Nature's domain with an observing eye and a sympathetic heart without being richly repaid. Once having developed trained powers of observation in wild places, an endless variety of form and of activity among living things appears on every hand.

Thus has been created a personal resource of interest and entertainment which is available in whatever part of the world one may roam. In exercising one's privileges as a citizen of the wild, the alert student may become intimately acquainted with the wonderful songs of the birds and the interesting habits and mental traits among the wild animals which often appear closely akin to our own experiences. In this wonderland the observer may follow many fascinating and marvelous occurrences, such as the changes which from a caterpillar produce a fairy-like butterfly, or from an unshapely underground grub a gorgeously colored beetle.

The marvels of the land are matched by those of the sea. Here giant whales, fishlike in form and habits, plow the deep, and being really mammals, suckle their young. In tropic waters the coral polyp patiently through the centuries secretes lime from the sea and builds up huge reefs which, eventually rising above the surface, become covered with vegetation and later make homes for men.

Among the interesting and often beautiful animals, although commonly abhorred by people through false conceptions concerning them, are the reptiles. Properly appreciated, they are wonderful examples of adaptation to special modes of life, and when viewed with an unprejudiced eye present forms and markings often of great beauty.

While the intimate relationship between man and the wild life with which he was surrounded in more primitive times has largely ceased to exist among civilized people, yet a strong economic connection is still maintained. For more than a thousand years whales have been pursued by men in ships for their oil and whalebone, and these great mammals of the sea are still being hunted. On the shores of the North Atlantic, particularly from the mouth of the St. Lawrence to Greenland, not only were whales once abundant, but many millions of harp and hooded seals abounded on the ice floes and gave occupation each spring to a fleet of vessels hunting them for their skins and oil.

The fur bearers, covering a wonderful variety of mammals from all parts of the world, yield a tribute each year valued at tens of millions of dollars to serve as garments or for ornamental purposes. Among our own prominent fur bearers may be mentioned the beavers, muskrats, sea otters, land otters, fishers, martens, minks, foxes, and others. For a long time after the colonization of the United States and Canada these areas were the source of a great part of the world's supply of furs. The foundations of many fortunes in the New World were laid in the fur trade, while famous organizations like the Hudson Bay Company and its rivals made their marks in history and had picturesque and romantic careers. The trappers and traders of the pioneer days were among the first explorers of great wilderness areas of this continent. Their stockaded trading stations formed the outposts of civilization across the continent to the shores of Oregon and north to the Arctic Coast, while the abundance of sea otters among the Aleutian Islands and along the northern Pacific Coast brought the Russians to occupy the shores of this continent south to California. The allotment of territory between the United States and Great Britain in western North America was, to a large extent, governed by the presence of fur animals and the activities of fur traders.

The buffalo herds of the western plains were estimated to number from thirty to sixty million animals. These disappeared before advancing civilization. Even as late as the early seventies trains passing through Kansas were stopped by vast herds of buffalo migrating to fresh pasture grounds. Until the engineers learned to respect the ways of these animals, attempts were sometimes made to force trains through the herds, which often resulted in the locomotives being ditched.

Having learned to make mental note of the images which are registered by his eyes, a person interested in Nature may here and there find remains of former animal life embedded in layers of rock or beds of clay. These serve as historical records, many of them running far back beyond the remote period when man first appeared on the earth. Among the fossils of these deposits have been found remains of a great variety of the mammals, birds, and reptiles with which this continent once teemed, but which through various causes have completely disappeared. Such finds open up another field for intensely interesting observation and study. Among the remains of former habitants of this continent are those of mammoths, elephants, camels, antelope nearly related to some of those now living in Africa, saber-toothed tigers, giant wolves, great bears far exceeding in size even the giant brown bears of Alaska, and many other kinds which no longer roam the earth.

As man developed from his most primitive stage, he began to exercise forethought to provide for a supply of food and other necessary material against the future by domesticating some of the useful wild animals about him. In this way have come our cattle, sheep, goats, horses, pigs, dogs, cats, and domestic fowls, all descendants from far distant wild ancestors, most of which are unknown today. It is an interesting fact that we are indebted to primitive man for almost all of our domesticated animals, since almost none of them have been brought under domestication within historic time. As civilized man continues to multiply and occupy all parts of the world, he carries with him his domestic animals and his cultivated plants, clears the forest, cultivates the ground, and destroys the wild life about him until its very existence on the earth is in danger.

Within comparatively recent years a sympathy with Nature and an appreciation of it have developed, and a multitude of people now feel a responsibility for the preservation of our birds and mammals and various other forms of harmless or useful wild life so that they may continue to lend the animation and joy of their presence. The earth would be a dreary and forlorn place if the innumerable birds and mammals and many other forms of wild life were destroyed. With the development of powers to see and to appreciate comes a feeling of sympathy and love for Nature, and it is hoped that the readers of this book may each and all have awakened an active desire to help save and perpetuate our wild life. In doing this

2

they will gain increased satisfaction in life for themselves and will help insure the joys of the wilderness for the people who will come in after years.

EDWARD WILLIAM NELSON
Chief, U. S. Bureau of Biological Survey

CHAPTER I

THE FROG

THANKS be to members of the frog family, for they invented arms and legs. The fish family is responsible for developing the backbone which has been found most useful through the centuries. It is more basic than arms and legs. Without the latter, however, there would be little use for pianos, and a jackrabbit would be a most ridiculous creature.

It is there in the water where it was shown that swimming could be best done by wiggling, that a jointed, flexible spine first came into being. It was when members of the fish tribe first crawled up on the land, probably because the ponds in which they were swimming had dried up, that they found that merely wiggling their tails was not enough. They needed some member that would help them to climb up the bank and to walk on the land.

As fishes they had fins, two on each side of the body. It was members of this frog tribe that first began to use these fins out of the water. That was back pretty well toward the beginning of things when all there was of America was a marsh from Philadelphia to Kansas City, hardly above the water and overrun with rank

1

ferns and pine trees. There had, up to that time, been
animal life only in the water.

Members of this frog tribe were the Columbuses of
their time. They struck out to plant their flag on new
lands. These fins, which had been satisfactory to their
ancestors, did not serve their purposes at all. They
kept trying to use them, however, for a million or so
years, and they improved with use as members are prone
to do. In the end legs were evolved. Like any other
invention that serves a good purpose their use steadily
spread. The first legs, however, employed certain basic
principles, as the patent lawyers would say, and those
principles are to be found in all the legs that have come
into being up to the present day. All are built on the
same general plan.

Out of this fish fin the frog folk worked a leg made in
three parts. There was the first single bone such as exists in your own upper arm. Then there was the second
part with two bones as in the forearm. Here was a device that would let the arm roll and twist. Then there
were the five fingers or toes at the end of the limb that
could be adapted to many uses, but whose chief business
was to get hold of things.

Every leg that has been used since the ancestors of
frogs came out of the water has employed the same principle, has used the same bone arrangement. In the flipper
of the seal is to be found these same three parts and the
five fingers. The bird grew feathers on its front legs and
used them for flying, but underneath these feathers in
this same bone structure the four fingers are to be found
in the last joint with the thumb sticking out, barb-like,

just at the bend. The stub foot of the elephant still shows five toenails. The whale which once lived on land and walked on four legs has flippers which have all the bones of the arm and hand and, underneath the skin, imbedded in the flesh where they cannot be seen at all from the outside, are the remnants of bones that were once hind legs.

The frog tribe, however, was not content with this single invention of arms and legs, great as it was. Like many modern inventors they went on and on to the creation of new devices. The heart of the fish was a two-chambered affair. They cut one of these chambers in two, putting in a valve, and the heart was greatly improved. Climbing out of the water they found that their old gill arrangement furnished a poor scheme for breathing and invented lungs and a nose to breathe through. Their breathing scheme was later improved upon, however, as frogs to this day have an awkward way of swallowing their air which would not serve marathon runners at all.

This creature that first came out of the water was not a frog as we know it today, but a less developed relative of the frog. It was more like the salamander which still exists in a modern world, an uncanny, undeveloped thing, often without eyes and living in damp caves or wells. The salamander is lizard-like in form. As a matter of fact, it is half fish, half reptile. It is the link between these two, one a water animal and the other a land animal.

The salamanders and the frogs are grouped together in an order which the scientists call amphibians. This

word means, literally, leaders of a double life. They are
so called because they are part water animals and part
land animals, half fish and half lizard. The frog is the
only amphibian of much importance.

It is remarkable that the frog should still exist in a

STEPS IN THE LIFE OF A FROG

modern world in a form so nearly like that in which it
appeared those millions of years ago when there lived
on the earth no beast nor bird nor bee, no moth nor butterfly, nor flowering plant. The frog then became and still
is an odd animal that lives part of its life in the water and
part of it on land, that is, independent of neither. Then
as now each recurring springtime heard, as an early evidence of its coming, the mellow call of the bullfrog. Then
as now that call was the song of a creature seeking its
mate. Then as now it was the forerunner of the appearance, in the waters of the early spring, of those clusters
of frog eggs imbedded in jelly or those long ribbons of
toad eggs that ·appear where there are quiet waters.
Then as now those eggs hatched with the developing

spring into the familiar tadpole, a water animal in every way, very much like the young of the fishes of these same waters.

These tadpoles, which steadily grew in size, were

AN IMMATURE FROG AND HIS OLDER BROTHER

little animals which breathed through gills just as do the fishes which swim about beneath the surface of the water. Some tadpoles were still water animals when the summer season drew to a close, still breathed through gills. Underneath their skins, however, if one should examine them carefully, might be seen the beginning of what were some day to be those new-found members of a developing animal, its four legs. Some tadpoles drive ahead and develop into frogs this first season. Those of the big bullfrogs, however, make the coming of autumn a midway station in their lives, go to the bottom of the stream, bury themselves in the mud, and sleep through the winter. Then, with the coming of another spring, they grow lustily, and crowd forth first their hind legs which they begin to use as paddles in swimming. A little later their forelegs come through the skin, and they begin to hang about the water's edge. They put their noses above that water and begin to gulp occasional mouthfuls of air. Their lungs take form. Presently they pull themselves upon the land or, if bullfrogs, upon the leaf of some pond lily. There they bask in the sun and breathe as do land animals. There the long tails of the tadpoles are gradually absorbed into their system. There they pass through, in a few weeks, that transformation which, at the beginning, required the cycling of many centuries. Then they become land animals. There they sit, fitted out with all those appliances which meant so much to the development of the animal world. There they sit, dumb, stupid creatures, compared with the more highly developed breeds of a modern time that have passed them by the wayside, yet offering their bodies as a living record of animal history of vast importance, written in the long ago.

The biggest of all the frogs is the American bullfrog. There are specimens of him that measure as much as seven inches from nose to tail. One may even find the tadpole of a bullfrog which is half a foot in length.

The bullfrog is not only the largest member of his family but he is also the member which is inclined to live most in the water. It is very rarely indeed that a bullfrog is found any considerable distance from the banks of ponds or streams. Only occasionally after long rains may one encounter such a frog, with the pioneer instinct, hopping methodically across country, probably on a pilgrimage to a new home.

The bullfrog likes quiet waters overhung with willows

and abounding in some such vegetation as pond lilies. In such waters in the summer time the bullfrog may be found sitting contentedly, partly in the water, partly out of it, always very alert and wide awake. He is wide awake because he is the possessor of a very great appetite and he is looking for food with which to gratify it. Much of this food he finds beneath the surface, for he is fondest of the little water creatures from which grow May flies and dragon flies. He also eats young crawfishes, other frogs and even bullfrogs of his own kind that are much smaller than himself. So also does he feed upon creatures of the air, upon grasshoppers and butterflies; and many a huntsman has missed taking home the bird that he has shot because it has fallen in the water and has been promptly swallowed by some great bullfrog. Even small ducklings, swimming on the surface, have been seen to disappear beneath it into the maw of a hungry bullfrog.

This greediness of the frog is equalled by the dangers that it faces through all its tadpole and adult life. There, in the bottom of the pond, the tadpole, because of its watchfulness and its speed, together with a knack of throwing out a smoke screen in which it escapes by dint of stirring up the silt with its tail, has taken part in many an adventure. There its tail may have been nipped off now and again by a hungry water beetle, but it has grown a new one in its place. As a frog it must watch eternally for the approach of some otter, skunk, or turtle, some crow or tall, blue heron, some stealthy snake which may steal upon it noiselessly through the waters, for it is a favorite food of all of these. Its refuge from most of them, to be sure, is deep down in the stream. It even buries itself in the mud at the bottom for safety, but, to its sorrow, it usually leaves its hind parts sticking out, thus giving the long-necked heron a clue to its whereabouts and a sure meal ticket.

Despite the fact that the bullfrog is a land animal which breathes with its lungs, it can go beneath the water and stay there.   When it goes down it closes its nostrils tight. It no longer uses its lungs, but gets its air through its skin which has some of the functions of a gill.   It is able to take the air out of the water as do a fish's gills.   The whale, which is an air-breathing water animal, can go under and stay for half an hour.   The bullfrog can do better than this.   It can go down for a week.   It cannot, however, stay down much longer.   This breathing through its skin is not entirely satisfactory.   While it is doing so, certain gases accumulate in its body.   They swell its body and eventually make it so light that it floats to the top whether or not its owner wants it to do so.

The bullfrog, sitting there on the lily pad, is a companionable sort of chap.   He is likely to look you over very alertly when you appear and to chug into the water if his suspicions are aroused.   If you move with deliberation, however, he will sit there and watch you.   If you dig in the earth, find a worm, and throw it on a near-by lily pad, he promptly pounces upon it, and you have gone far toward winning his friendship.   If you repeat this for a few days he is won, will allow you to take him up in your hands, and will become quite friendly.

The green frog is given to screaming quite distressingly

when, in fright, it plunges into the water. The bullfrog also screams in a manner almost childlike when it is caught by some one of its enemies. It screams in a high note with its mouth wide open. This note is quite different from its "jug o' rum, jug o' rum, jug o' rum" call of the mating season, during which time it is the heavy bass of the swamps.

Those members of the frog family that are quite independent of the water, that live in it only through the spring breeding season, whose favorite place of abode is the flower garden or vegetable garden cultivated by man, those amphibians that present an ugly and wart-covered back to the world, are nearly everywhere known as toads.

These toads are the best known of all the group. They are so useful to man that they may almost be termed domestic animals. They live almost exclusively on insects, and insects are likely to be injurious to gardens. Their value to man is recognized in France where they are bought by the peasants and released in their gardens. Careful students of their feeding habits estimate that a single frog in one's vegetable garden may be worth twenty dollars to him in a single season. Insects are man's greatest enemies, are growing in power upon the earth, are threatening some day to take possession of it. Since the toad eats insects, it is one of the elements that tends to hold back their increasing menace. This service of the toad is not generally appreciated and, though generally there is a kindly feeling toward them, they are not given the protection and encouragement that they deserve.

The toad likes man's garden better than the wild, open

field because certain of its worst enemies, for instance, the owl, is less likely to come there in its hunting. It seeks damp places in which to sleep through the heat of the day, and retreats beneath board walks or platforms around well curbings when it is wet. This is because the only water it gets is that which it absorbs through its skin, for it never takes a drink through its mouth.

Then, as the cool of the evening comes on and insect life is busiest in the garden, the toad hops forth for its dinner, bringing with it a surprisingly large appetite. It is not at all particular about what it eats as long as it is something that lives and moves. Its idea is to capture any tiny creature that moves. If the insects of a garden would sit still, the toad might starve to death with food all around. It preys upon moving objects only. A favorite joke on the toads or bullfrogs is based on their greed and readiness to seize any moving object. If leaden buckshot are rolled near their noses they will swallow them, often-times swallowing so many of them that they may be weighted down so that they can no longer hop. Certain caterpillars know that the toad will attack only moving objects, and so, when a toad appears, they roll up in balls, play dead, and are safe.

In catching insects the tongue of the toad is its greatest help. It is an odd sort of tongue, with glue on the end of it, that works on a hinge at the tip of the jaw. When the toad gets near an insect it flips out this tongue to a dis-tance of about two inches and catches its victim. It may sit by the kitchen door by the hour and capture flies and mosquitoes. In the garden it snaps up grasshoppers, cutworms, beetles, caterpillars, potato bugs, spiders, ants, even the sluggish and slimy garden slug that comes out by night and eats the young leaves of the lettuce. It occasionally devours some useful insect like the ladybug beetle, but nearly everything it eats is injurious to the garden. Often, when a garden promises to be a failure because of insects, it may be saved by putting a dozen toads in it.

The warty back of the toad is a method of disguise. It is made up to play the part of a stone or clod of dirt that it may not readily be seen by its enemies. Its color, also, may be gray like the ground or green like the grass.

The game of keeping alive has been a desperate one for the toad. It is a clumsy creature, a product of an age when animals were not highly developed and by no means as physically fit as modern mammals. It falls an easy victim to almost any of them that see it. Its safety depends largely on its not being seen. So it hides under some such place as the doorstep and makes itself hard to see even when it is abroad.

Then it has another trick for its own protection. When it is captured by an enemy it sends forth from glands on its back an acrid secretion that is exceedingly disagreeable. A dog, for instance, if it is a very young dog and without experience, may take a toad into its mouth. The toad immediately ejects the secretion through its skin, and the dog is made so sick that it drops its victim and shows every sign of distress, and is unlikely again to make a similar mistake. A wise dog may play about, threaten and bark at a toad, but will not bite it.

There is a very old belief that these secretions of the

toad cause its own warts and also that they will cause warts to develop on the hands of those who pick up the toad. This is in nowise true. One does not get warts from handling toads, for these little creatures are quite harmless and do not even take the trouble to put forth these offensive excretions unless they are frightened.

The toad has yet another method of keeping itself going in a world full of enemies. It breeds multitudes of little ones, realizing that the great majority of them are to be sacrificed before they come of age, and that only one now and then will grow to be a big toad and return to the pond to lay eggs for another generation.

Those that do survive become water animals again for two months in the spring. It is then that they sing their plaintive and melodious mating song on the water's edge, for the toad has the sweetest voice of them all. It is then that they lay their strings of eggs in the water, a single individual producing thousands of eggs. Presently the eggs become little fishlike pollywogs. These pollywogs are so innumerable at places the water may be black with them. In a few weeks their hind legs appear, with feet not webbed as are the bullfrog's however; and later the front legs appear. Now the pollywog begins to come to the surface and gulp mouthfuls of air. It is no longer a water animal, but a land animal. It would drown if you held it under water too long.

The pollywog crawls out on the bank, it absorbs its tail, expands its lungs, and practices a bit at hopping. It is now a toad, though a very tiny one. There are millions of others just like it all around. Presently a day comes, a day of rain, which is the best time for it to start an independent life of its own as a land animal. The young toad and all the rest of its baby associates start hopping inland away from the pool. They have set out on their great pilgrimage. It is like a Light Brigade in a desperate charge. Many will die, but some few will get through. Larger creatures, man and his vehicles, trample them to death. Some get through to the woods, the fields, the gardens, and begin that game of hide and seek with their enemies which, for a few, will last the two or three years that it takes to grow—a desperate existence for so mild a creature.

Outstanding as a type of frog of an entirely different mode of life is the green tree frog, but one of many varieties of frogs that are given to the habit of climbing. These frogs live like squirrels in the trees, feeding on insects. The green tree frog is one of the slenderest and most active of them all. It is said to be able to leap eight feet and is pretty likely to get a fly at a distance of four feet. It spends the summer in the trees, sleeps through the winter in a rotten log or in the ground at the tree's roots. Then in the spring, that busy season in the ponds, the green tree frog, like the toad, takes to the water for a brief period, during which it lays its eggs and thus provides for the new generation.

There are many varieties of bullfrogs, toads, tree toads, all quaint and interesting, mild and harmless—all inclined to be friendly to man, easily tamed, musical and entertaining. These queer animals, these amphibians, at home both in water and on land, these survivors of a time when the world was young, are among the most interesting individuals in the animal kingdom.

QUESTIONS

1. The steam engine began when boiling water lifted the lid on a tea kettle. The electric dynamo goes back to the time when Benjamin Franklin flew a kite in a storm. What do you conclude as to the origin of other things about you? Of all things? Where did corn originate? tomatoes?

2. Where did arms and legs get their start? backbones? Can you imagine the sort of animals that lived in the world before that time? How long do you suppose it took to develop backbones or legs? Is the principle of this early leg still in use?

3. What are the animals called that are half water, and half land animals? Which of them may be seen in your neighborhood?

4. What is the difference between a frog and a toad? Can you tell the eggs of a frog from those of a toad? How many members of this class will promise when summer comes to get both frog's and toad's eggs and keep them in water while they hatch, and change from tadpoles to frogs? Do both become frogs?

5. What does the bullfrog eat? What eats it? How would you proceed to make friends with a bullfrog?

6. Where do toads live? What do they eat? Do they have jewels in their heads? Do they make warts on your hands? How can man make practical use of toads?

7. The toad is so clumsy and helpless that you would think it would be quickly wiped out by its enemies. How does it escape?

8. Does the toad raise a large family? Why? Describe the inland migration of young toads.

9. Do frogs live in trees? When and where would you expect to find them in trees?

10. Which are the older, men or frogs? frogs or fishes? In your mind can you picture that time in the world when frogs came out of the water and were the most advanced of all living things?

Chapter II

THE DOMESTIC CAT

S O commonplace is Tabby, asleep there on the rug, that people rarely stop seriously to consider her, and so fail to realize that she is one of the most remarkable creatures in all the world and that there is back of her a historical romance which spans more than four thousand years.

Her loving mistress is not likely to know, for example, that Tabby was once a god. She probably does not even realize that Tabby is of Egyptian descent. She may not be aware of a fact of this present day, that Tabby leads a double life, that she is the Dr. Jekyll and Mr. Hyde among man's animal associates, that she is an outlaw in the eyes of many organizations interested in bird protection. Her mistress may not know that, wrapped up in the fluffy skin of this, her pet, is one of the fittest machines in bone and muscle that exists in all the world. She probably has not learned that Tabby is just now menacing the welfare of the great continent of North America by contributing a liberal share toward upsetting the balance of nature.

There are remarkable facts back of this purring, permanent guest in the house of man—facts that are overlooked because cats are so commonplace that one never stops to give them thought. Let us start at the beginning to try properly to appraise this most familiar of animals.

In the first place, historically, there is no mention of cats in the literature of ancient Greece or Rome, or of the Near East in Biblical days, or of China or of any other country down to the Christian era with the single exception of Egypt. In the art of Egypt, two thousand years before Christ, there began to appear pictures of the cat associated with man. Here it made its first bow as a domestic or household animal.

These pictures were of a cat for all the world like the common, yellow or gray, short-haired tabbies of today. Such cats still exist unchanged in the wilds of Nubia and Sudan. One of them, captured four thousand years ago and brought up in the hut of an early grain farmer of the Nile, would cuddle into the crook of its master's arm and purr contentedly. It would raise a family of cats like itself, and these would hang around the place and come to fit into the scheme of things. It was a part of the nature of these cats to be easy to tame and to stay tamed. It is a peculiar thing that other cats would not submit to being thus tamed. So it came to pass that the Egyptian cat became man's household companion over the greater part of the world.

These cats proved useful in the valley of the Nile, which was a grain country and therefore infested with mice. But, more than this, they came to be worshiped. Later the great city of Bubastis was built and dedicated to cat worship, and it is recorded that seven hundred thousand people made pilgrimages to that city in a single year. On one occasion the Egyptians and the Persians were at war, and the Persians had been defeated in battle. They returned to the fray bearing live cats as shields, and the Egyptians surrendered their city rather than endanger the cats.

It was after the time of Cæsar that these Egyptian cats were introduced into Italy, and the Romans, during their occupation of the British Isles, first brought them to England. They were not worshiped in Europe, but were, instead, often regarded as creatures of evil. Even today it is supposed to be bad luck for a black cat to cross one's path. Finally, when America was settled by Europeans, the descendants of these Egyptian cats crossed the Atlantic.

In the East, in Asia, there is another related type of cat which has been domesticated and which is recently finding its way into Western countries. Angora and Persian cats are the best known specimens of this type and are now frequently seen in America. All of these long-haired cats are Asiatic. They come from a wild strain of their kind that has yielded itself to domestication. Then, there is an occasional strange creature from some isolated region. There is, for example, the Siamese cat which may be light orange in color with blue eyes, or the blue cats of Malta, or the Chinese cat with the short, hooked tail. These cats are occasionally met with over here, and there is no question as to their origin.

But the mass of Western domestic cats are of Egyptian blood. There is a certain peculiar thing about them that distinguishes them from other cats of other continents. The tough skin on the bottoms of their feet is black, and this blackness extends up into the hair.   This is not true of other cats.   It is the racial Egyptian trait brought down through the centuries.   By examining the bottoms of its feet one may tell if his own cat is an Egyptian.

In Europe these house cats have mixed to a degree with the wild cats and have been changed somewhat, but on this side of Asia no other breed has ever been successfully domesticated.   In Europe there has always been a striped wild cat, not so different from these house cats, but its nature has been such that it could not be tamed.   In America the native bob-tail wild cat has not been domesticated.   There is something in all these other cats that seems to keep them wild.

There are many cases of this sort among animals, for example, the zebra is very close kin to the donkey.   The zebra has within it a wild spirit that refuses to be bent to the will of man.   The donkey, on the contrary, lends its back to man's burdens the world around.

This yielding to the will of man, this willingness of a wild animal to become domesticated, has worked out very greatly to the benefit of the Egyptian cats.   In America the wild cats have been driven back and back into the waste lands.   Their numbers have steadily decreased. Every man's hand is turned against them.   Advance is being made steadily toward the time when there will be no more wild cats.

But this descendant of the domesticated Egyptian cat is faring quite differently.   It has found comfortable lodging in the houses of the settlers and has spread and multiplied with them.   It has fared afield and captured its fresh meat if it chose, but when it saw fit, it has coaxed a living from the people with whom it dwelt by the simple process of rubbing an arched back against their legs and uttering an occasional plaintive "meow."

While the numbers of the native wild cat have decreased, those of the domestic cat have multiplied many times. Most people have cats in their homes. All the towns and cities have large populations of cats—alley cats, ash barrel cats—that have no masters, but forage livings for themselves. Wherever there are woods or waste land, there are to be found the descendants of domestic cats that have gone back to the wild. A New England farm may have been abandoned and the cats left to shift for themselves. A family living in a village or town may have gone away for a summer vacation and left the cats to find their own food. Every summer-cottage settlement finds in the autumn that many cats have been left homeless. People, too soft hearted to kill their cats that have grown overnumerous, often take them into the country and turn them loose. All such cats go back to the wild and breed new generations of their kind.

So it is that there are in the United States many many more of these cats whose ancestors came from Egypt than there ever were wild cats in our wildernesses. Students of the cat population estimate that in the United States there may be as many as fifty million. They are to be found far back in the mountains scores of miles from human habitations. Go into the woods almost anywhere in the winter after a fall of snow and examine the tracks of wild creatures that have prowled abroad. You are likely to find that half of the tracks left in this fresh snow are cat tracks, the tracks of creatures of this domestic breed. So it happens that this introduced species has come to occupy an almost dominant place in the wild life of a nation.

It is just here that Tabby begins to perform that outstanding feat in her whole career, that of upsetting the balance of nature. It is just here that Tabby, by assuming an overambitious rôle, by becoming stronger than it

THE EXTINCTION OF INSECT DESTROYING BIRDS IS THE GREATEST DANGER
THREATENED BY CATS

was conceived that she should become, has placed herself in the position where she may array the man creature, who has always been her protector, against her and set him on a campaign for her extermination.

The cat is becoming a menace to many of the wild creatures that make up a part of the animal life of the nation—creatures many of which are helpful to man or at least contribute greatly to his pleasure. The cat threatens the very existence of the bird life of the nation. It is a menace to the existence of such creatures as rabbits, squirrels, moles, and shrews. It is a destroyer even of the poultry of the barn-yard.

The fifty million cats in the United States must, like everything else, eat to live. They are not particular about their food as long as it is fresh meat, but all of them eat fresh meat. They are that kind of animal. They sometimes eat insects, are fond of lizards, and may learn to catch fish despite their dislike for water. But, above all, they are fond of the flesh of birds.

The chicken-killing cat—a numerous tribe—is on a par with the sheep-killing dog. Pigeons and doves are still more attractive and more liable to attack. Colonies of gulls and terns on islands along the coasts have been known to have been wiped out by a few cats.

Some cats are "ratters" and some are not. The defense of the cat is that it keeps the rats and mice away, but there is much testimony from experts to the effect that cats and rats may live quite happily together, neither disturbing the other. Destroying rats has become quite a science of late and the best experts combat them, not with cats, but by rat-proofing buildings, putting out poison, and by setting traps.

And, finally, the great arraignment of the cat is as a bird destroyer. One observer reports that robins, building their nests in his orchard in places easy to find and to reach, had been wiped out—a score of them—by one mother cat. Another had a wooded yard which was inhabited by eighteen warblers, two wrens, two woodpeckers, and four unidentified birds, and he saw them gradually disappear as victims of a single cat owned by a neighbor.

One owner proudly boasted of the prowess of his cat which, though fed at home, brought in twelve birds in two days.

But this sort of thing is not the worst. The mild-seeming cat which plays with grandma's ball of yarn with the handiest pair of paws in the animal world, which daintily munches blades of grass on the lawn on a summer evening for the aid of its digestion, which sleeps by day in the sun, is a different creature by night. By day she

THE CAT FAMILY ALONE CARRIES ITS OFFSPRING IN THIS MANNER

is mild and lazy, soft and pliant. By night her body tightens itself like a bowstring. No other animal has a frame so agile and flexible that it can spring into the air and catch a bird on the wing. No other animal can fall from such heights without suffering injury. No other kind of animal has claws that will cut like a knife or grip like the iceman's tongs, yet which retire when not needed into the pads of its feet and are practically invisible. No other creature in all the world has stronger teeth, specially made for tearing flesh, operated by stronger muscles. No other creature, except possibly the owl, has such marvelous, night-seeing eyes, eyes in which the pupils may be so dilated at night that they cover the whole front of the eyes, thus letting in whatever dim light there is, and which may be contracted to a narrow slit in the noonday sun.

The cat is a member of the carnivorous, or flesh-eating order of animals. Another great branch of flesh-eaters is the dog family. The dogs are more simply organized, specialize in running after their game, never think of climbing a tree, nor tearing an enemy with their claws.

Then there are the fur bearers of considerable variety of form and habit that are flesh eaters—some lithe and agile like the weasel and others clumsy fellows like the bear.

Few of these compare with cats in bodily activity, in prowess in the hunt, in fighting ability when cornered. As animals are classified by the zoölogists, a family is divided into genera. The Canadian lynx with tassels on its ears represents a genus of the family that is different from that of the domestic cat. The great cougar, or Rocky Mountain lion, which in the west kills colts, calves, and deer; the dappled jaguar of South America, and the leopard of Africa, which wears a spotted skin to look like splotches of sun and shade, in the trees where it lives, are all cats. So is the lion of Africa, wrongly styled the king of beasts because another cat, the tiger, is bigger, fifty pounds heavier, and has mastered the lion on many occasions, notably in the days when such creatures were matched in battle in Roman arenas. They are all fellow members of the same genus *Felis* to which the domestic cat belongs.

Thus does it come to pass that this cat finds its place in the animal kingdom by the following system of classification:

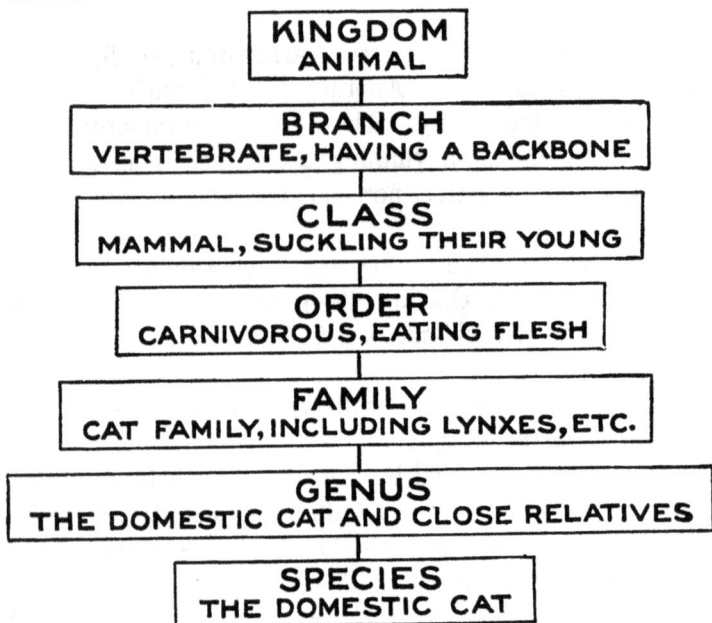

```
KINGDOM
ANIMAL

BRANCH
VERTEBRATE, HAVING A BACKBONE

CLASS
MAMMAL, SUCKLING THEIR YOUNG

ORDER
CARNIVOROUS, EATING FLESH

FAMILY
CAT FAMILY, INCLUDING LYNXES, ETC.

GENUS
THE DOMESTIC CAT AND CLOSE RELATIVES

SPECIES
THE DOMESTIC CAT
```

Of all these cats, only one has the knack of adapting itself to life in the house of man. That life has in no way changed the night nature of the house cat. When its master is asleep it steals forth to murder whatever creature it encounters that may be weaker and less physically fit than itself. With all its prowess it never battles another animal that is anything near a match for it unless cornered and forced to do so. In fact, its favorite victims are entirely helpless. Its choice of them all is the defenseless mother bird in her nest at night warming the eggs that she hopes to hatch into other little birds. The stealthy cat climbs in the darkness to such a nest, surprises the mother while she sleeps, clutches her in a death grip, wrecks the nest, and skulks away in search of other victims. Her mistress little suspects these crimes of the night. Madam Cat lives luxuriously and lazily during the season when young birds are helplessly trying their first wings. She slays even when her stomach is so full that she cannot eat.

Before the coming of white men wild life in America had struck a certain balance. There were enough wild cats and other creatures that prey to keep the birds and rodents from becoming too plentiful. There were not enough of them to destroy this wild life, for there were larger beasts, in turn, to prey upon them. The birds and rodents, in turn, kept insects from becoming too plentiful and destroying plants. Nature, left to herself, had established a balance.

But man introduced a new kind of cat, bred it in great numbers, killed the wolves and lynxes that might have fed upon it, created and is helping to maintain an unnatural situation—that of too many cats in proportion to the birds. So is the balance of nature upset. So is the extermination of the birds threatened. So may the insects, man's greatest enemy, become too plentiful.

Few people have ever stopped to think of the cat as a public nuisance—to realize that it is becoming a menace. The State of Massachusetts and other States have done so and have taken corrective measures.   The United States Government, through its Biological Survey, has studied the problem and pronounced the homeless cat an outlaw that should be fought.   Many individuals and societies have investigated the situation and agree in two conclusions—that ownerless cats should be killed and that owned cats should be kept in control.   Tabby who gives pleasure around the house in the daytime should not be turned loose outside at night to become a deadly murderer of creatures more important to man than she.   She should be kept inside where, if she goes hunting, she will render a proper service by killing rats.   If turned out at all, her owner should wait until after she has had her breakfast in the morning.   And even then she should be kept inside during the spring breeding season for the birds.   The ownership of cats should carry with it some of the responsibilities that are attached to the ownership of certain other domestic animals.

QUESTIONS

1. Where do cats come from? How did they get to America? Where did the long-haired cats come from?   Examine the cats in your own home, and any other cats that you may see, and report to the class as to whether they are Egyptians or Asiatics.
2. Are there more native cats than cats of imported breeds in America? Do you often see these imported cats that have no masters? Did your family ever turn out any cats to go wild?
3. What is the favorite food of cats? Watch your own cat and try to find out which it eats more of, rats or birds. Watch the nests of such birds as robins and see if the young ones live to grow up. If not, do you think cats were to blame?
4. What does Tabby do through the night while you are asleep?
5. Why are cats called "carnivorous" animals? Name some other carnivorous animals.

6. To what "order" does the cat belong? what "class"? what "branch"?
7. What do you know that shows that the cat is more active than other animals?
8. What is meant by the balance of Nature? How has it been upset? Is man to blame?
9. The government does many things for the benefit of its citizens. Did you know before that it had a bureau for the study of animals?
10. Does the community in which you live do anything to keep down the number of ownerless cats? Maybe even the grown men have never thought of the harm they do. Perhaps it is a task for the boys and girls to get people to understand this problem.
11. Just what do you think people in your community should do about the ownerless cats? What should they do with their pet cats to keep them from eating the young birds?

CHAPTER III

THE SHARK

TWO gladiators of the deep were fighting it out one summer afternoon off the coast of North Carolina. A very fortunate group of travelers aboard a lazy coastwise steamer happened along at just the right time to witness the combat. The fighters were none other than tiger sharks, those cruisers of many seas, who are believed the world round to be much given to the practice of eating human beings.

Upon this occasion, however, two of these tiger sharks, each some twelve feet in length, were in a death grapple with each other. Oddly, armed as they were with huge mouths full of teeth built for cutting and tearing, this was a bloodless battle. Those who watched it were surprised to find that these vicious fish did not attack each other with their teeth. The weapons which they used were tails. They beat the water frantically as each played to get in a position where he could deal the other a blow with his tail, could strike as with a club, could deliver a knockout blow. They did, in fact, fight to a finish, until one had beaten the other to death with its tail. Those on board the lazy steamer who had stopped to witness the conflict saw the beaten shark turn on his back as fish do when dead. They made preparation for taking the carcass aboard ship. Before this could be accomplished, however, the victor returned, seized his victim in his mighty jaws, and bore him away.

These sharks are cannibals; they eat one another.

A COMBAT OF TIGER SHARKS

Proof of this assertion was obtained in Charleston Harbor not long ago when a school of them appeared and swam about a ship at anchor. They seemed to be very hungry female sharks, and no sooner was a hook baited and thrown overboard than one of them seized it. She was dragged aboard and found to measure some nine feet in length. A second shark of similar size was immediately caught. While this shark was hanging by the tail from the boom, its head just out of the water, a third member of the school, twelve feet in length, leaped out of the water to get at it, ripped it open, and began to devour huge mouthfuls. While it did so the captain began shooting it with a revolver. This seemed merely to infuriate the beast, which returned to the attack, until it was finally hit in a vital spot and killed.

Another evidence of the bloodthirsty nature of these wolves of the sea was presented off the coast of Florida, where a number of them were seen in pursuit of a porpoise. This porpoise probably had been injured in some way, for sharks are given to attacks on disabled porpoises. They hunt big game at sea, as the wolves used to hunt buffalo on the plains; they follow the herds, but attack only the individuals that are disabled. This group of sharks drove the injured porpoise into shallow water and there tore it to pieces and devoured it.

Sharks sometimes attack human beings swimming in the water. They bite off an arm or a leg and devour it. It might even happen that a whole body is eaten by sharks. Usually such an attack is made where the swimmer is in distress, is suffering from cramps, or is in a drowning condition. These sharks, vicious as they seem, are great cowards and, under normal conditions, give man a wide berth.

The stories of sharks attacking human beings are greatly exaggerated. The waters along the coasts of the United States always have sharks in them, yet people swim in them all the time without being molested. In the waters of the Caribbean Sea, around Cuba, Santo Domingo, and Jamaica, there is a great abundance of many species of sharks, yet the natives swarm in these waters.

In the South Seas shark fishermen throw their lines in among the bathers and catch their prizes. Despite the abundance of sharks it is difficult for a careful investigator traveling around the world to pick up a dozen proved cases, in a generation, of sharks having attacked human beings. They do attack occasionally, as lightning occasionally strikes on land, but seldom more frequently. Stories of attacks by sharks are usually not true.

Sharks are the largest of all fishes. From the standpoint of size they are to fishes what elephants are to land animals. There was a time on land when huge creatures, many times as large as any now alive, stalked through the jungles. The elephant is the sole survivor of that class of animals. Likewise, there was a time when the ocean was inhabited by fishes many times as large as those that now exist. The shark is almost the only survivor of this early order of fishes.

The largest of these sharks, and therefore the largest fish in the world, is the whale shark, wearer of a polka-dot dress, child of the tropics, whose natural home is the Indian Ocean. Strays from these herds, however, sometimes drift around the Cape of Good Hope, roll along up the Atlantic Ocean, and flounder into American coastal waters down Florida way, where fishermen are plentiful. In these waters two specimens have been captured in recent years.

This is a huge, clumsy monster forty or fifty feet long, with a blunt nose, and there may be whale sharks one hundred feet long. It is not a hunter fish like most of its fellows, has none of their ripping and tearing teeth, but whole fields of small teeth not larger than

4

buckwheat, used for purposes of grinding such food as small shell-fish.

Cousin to the whale shark, likewise dressed in polka-dots, is the basking shark, a huge, lumbering creature, given to sunning itself on the surface of the ocean, nearly as long as a whale shark and as thick as a man is tall. This big fish comes out of the north and may be seen rolling in the waves as far south as Santa Catalina on the one coast, or Virginia on the other. This, also, is not a fighting shark. It is easily harpooned and often falls victim to whalers who are chiefly interested in the oil that may be secured from its liver.

Next in size to the basking shark is that individual most likely to attack man, the great white shark, thirty-five feet in length and with a mouth of such size that it might easily gulp down a grown man. Proof of this assertion is the fact that, in the stomach of one of these great white sharks once killed off the coast of California, was found the body of a young sea lion weighing more than a hundred pounds.

These great white sharks are undoubtedly the most dangerous of them all. They are vicious beasts of prey armed with row after row of pointed, cutting teeth that might tear a horse to pieces. They are of a species that is not very abundant, but which exists all around the world, and is sometimes seen as far north as the New England coast. They roam the ocean without fixed abode, and a single individual might travel over much of the world.

It is probably the great white shark that is responsible for most of the occasional human deaths from big fish, but in this case also there is little proof. They loaf along the Atlantic Coast of the United States in the summer time, for instance, but almost never attack bathers.

There is the one case in recent years of such attacks, that of 1916, when along the coast of New Jersey four persons lost their lives. The marauder in this case is believed to have been a renegade white shark, one individual which was responsible for all the attacks. Such a shark is in the same class as the occasional elephant that goes "must," or the occasional shepherd dog that becomes a sheep-killing dog. It is an outlaw, and the others of its kind are very unlikely to follow its example.

Spotted tiger sharks swimming in schools, vicious, long-toothed fellows, sometimes thirty feet in length, are to be found in all the oceans. So, also, are the blue sharks, which glide almost invisibly through the water. So is the smaller sand shark, five feet long, built on rakish lines, sharp-toothed, said by fishermen to have a wicked eye. This shark is quite capable of snapping off an arm or leg if it sees fit. The dogfish, a yet smaller shark, traveling in packs, is hated by fishermen because he preys upon fish on lines or in nets, lobsters in pots, and thus interferes with their catches.

Another odd fish found in all the oceans is the thresher shark, which may be twenty feet long and which has a tail one third the length of its body, but extending fantastically off at an angle. It fights with this tail, and is said to deal whales resounding smacks with it when they come up to breathe. Its real purpose, however, seems to be to beat up the water and confuse the fish that it is trying to catch for its dinner. Another strange member of this family is the hammerhead shark, so named from the shape of its clumsy head, which looks not unlike a sledge hammer to which the slim body, ten feet long, plays the rôle of handle.   It is abundant around New York Harbor.

These sharks are kings of the fishes.   But there is a more familiar creature of the waters, the ordinary porpoise, which one is likely to see leaping above the waves as he sails out of any harbor in America, and which does

A HAMMERHEAD SHARK

not recognize the mastery of the shark.   The porpoise, one very naturally concludes, is itself a contender with the shark for first place in size among the fishes.   It happens, however, that there is no contest between the porpoise and the shark, because the porpoise is not a fish at all. It is a member of the whale family, a little cousin of the whale, as it were, and whales are not fishes.   They do not breathe through gills, they do not hatch their young from eggs and desert them. They are mammals. Their young are well developed when born. They keep them about, take care of them, nourish them with milk just as does the cow. This fact that they provide milk for their young puts them in the mammal class. Whales are much more closely related to cows than they are to sharks.

As fishes go, the shark, though the largest, is one of the least highly developed. The shark is an old form, is a survival of an earlier age in the animal history of the world, an age in which, as clearly shown by fossil remains in rocks, the sharks were many times bigger than they are today. The shark does not even possess a fully developed backbone. Its backbone is merely a cartilage, a piece of gristle. It has not in it the degree of stiffening which warrants its being properly ranked as bone. It can be cut with a knife. It does not have the ribs of other fishes to inclose its inner organs. Its outside muscles, having no bones to tie to, fasten themselves to its hide. The shark is a lower order, in the course of developing a backbone, but has not quite completed the job.

It may be said, even at the risk of reputation, in getting an understanding of the animal kingdom, that there are two great divisions of that kingdom, as follows:

(1) Those animals that have backbones (vertebrates), for example, fishes, birds, frogs.

(2) Those animals that have no backbones (invertebrates), for example, oysters, crawfish, grasshoppers.

All of these, even the insects, are animals. People are likely to think of the word "animals" as applying only to the familiar four-footed creatures about them. Properly speaking, an animal is any member of the Animal Kingdom, such as birds, snakes, fishes, oysters. The animals that we have so far studied have all been of that most important class known as mammals— animals that feed their young on milk. This class has been divided into orders, families, genera, and species. To be able properly to place the fishes, however, we have to go further back than the mammals. We must go back to the vertebrates, because the fishes are not mammals, but, like the mammals, have backbones and, therefore, are vertebrates. They make up one of the five great classes into which all animals with backbones are divided. Thus, we have the diagram as follows:

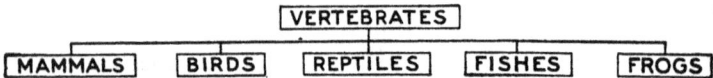

```
                    ┌─────────────┐
                    │ VERTEBRATES │
                    └─────────────┘
  ┌──────────┬──────────┼──────────┬──────────┐
┌────────┐ ┌───────┐ ┌──────────┐ ┌────────┐ ┌───────┐
│MAMMALS │ │ BIRDS │ │ REPTILES │ │ FISHES │ │ FROGS │
└────────┘ └───────┘ └──────────┘ └────────┘ └───────┘
```

The animals without backbones came first in the development of the world. Those with backbones are the higher order and developed later. They developed from those that had no backbone.

The world was originally so hot that there could be no life upon it. The theory is that vegetable life developed first, since plants live on minerals, and these were present. Animals live on plants or on other animals that live on plants and so could not have developed until there was plant life. It is believed that the first plant life began in the water as soon as it was cool enough to allow that life to develop. The first animal life also began in the water, feeding upon this early plant life. There are still plants and animals in the ocean that are so tiny and so simply organized as to seem very like what must have been the original forms when life first appeared on the earth.

In the sand in the bottom of Chesapeake Bay is to be found an odd creature about two inches long, called a lancelet. In it is a backbone that is just taking form. This creature is halfway between a clam, which has no backbone, and a fish, which has a backbone. It is a living evidence of the route by which the animals with backbones developed. It is the connecting link between the two branches of the Animal Kingdom. It has a backbone near the beginning, while the shark has a backbone which is much farther advanced and which has become perfected in the mammals.

Animals with backbones first became highly developed in the water. Fishes are older in this world than are land animals. A minute, jelly-like organism endowed with life came into being. Then followed a long series of strange transformations always increasingly complicated until the beginning of a backbone appeared in lancelet-like animals.

At the end of it all, as a great monument of accomplishment, stands the backbone, the master structure of the ages. The development of a backbone is the real foundation for the development of the higher animals as we know them today. It is through the fishes that we have attained this development; and the fish stands out, therefore, as a great contributor to the development of the higher animals with backbones, and therefore man.

As the fish took form, it developed fins on various parts

of its body, as, for instance, at the end of its tail, along the ridge of its back and along its sides. Originally there was a ruffle of fin running along the sides of the fish. This later divided and there became a fin forward and one aft on each side of the fish. These four fins were arranged in pairs and were the beginning from which have developed the four legs of the higher animals that today inhabit the earth.

The fish used these fins in its swimming and in maintaining its position in the water. Later some of them began to use them to hold on to various objects. There are the gobies today, for instance, which have a sucking disc on their fins, by means of which they fasten themselves to rocks. There is also a variety of perch which uses its fins to climb up the banks of streams and is even said to climb slanting tree trunks with them. The so-called frog fish can use its fins for walking and takes refuge on land for protection from its water enemies.

Certain fish have developed the ability to travel overland when one stream or pond dries up to find another. They develop their fins to do this.

It was in this way that the first backboned creatures came out of the water and took up their abode on land and developed into the backboned animals of today. It was the fins of these fishes that became the legs and arms of man. Next to a backbone these members may be considered as developments that have been of greatest use to a growing world.

In the study of the shark we encounter the remora, a fish a foot or two long, which furnishes one of the queerest examples in all nature of a part of the body being put to a use for which it was not at first intended. In this case it is the fin on the back of the remora that serves its odd purpose. Ages ago this little fish formed the habit of swimming underneath the shark, probably for protection. Gradually its back fin flattened out that it might stick closer to its protector. Then by degrees this flattened fin became a suction disc. The remora, armed with this suction disc, now swims beneath the shark, presses this disc against the shark's abdomen, and thus fastens itself on to the shark. Thus can it ride about in safety without so much as a wiggle of its tail, for other big fishes that might otherwise devour it, do not dare to come near the shark.

Aside from the difference between the backbone and ribs of sharks and those of other fishes, the shark has, also, different kinds of gills. It has five gill slits on the sides of the neck through which it breathes, while other fishes have but one opening. The five openings are an old-fashioned arrangement in the fish world and go to prove again that the shark is an ancient form not nearly so highly developed as the others. On the sides of the heads of these other fishes, for instance, may be seen the marks of what were once these five openings, used before the new sort of breathing arrangement came into use.

QUESTIONS

1. What impression do you have of the nature of sharks? Do they seem to be pleasant companions to each other?
2. To what extent do sharks attack human beings? What does the book say about stories of sharks attacking people?
3. How do sharks rank among the fishes as to size? If the shark is not as large as a whale, how can it be the largest of fishes?

4. Describe some of the cousins of the shark family. Which are the most dangerous?

5. Is the porpoise related to the shark? Where does it fit into the animal world?

6. The frogs were among the first land animals to develop and are of a lower order than, for instance, the cats. The ungainly camel is of an old order which has nearly passed and is less highly organized than the horse. What can you say in this connection of the relation of the shark to other fishes?

7. Describe the backbone of a shark. What can you tell about shark ribs?

8. Are sharks animals? Are birds animals? snakes? cockroaches?

9. What are the five classes into which animals with backbones are divided?

10. How did animal life get started in the world?

11. What animal developed the backbone? Which do you think the more important, the development of the backbone or the development of the steam engine?

12. Aside from its invention of the backbone, what other important contribution did the fish make to the development of animal structure? What was the beginning of the human hand?

13. The lancelot, buried in the mud for ages, is a great deal like Langley's airplane, put away in the National Museum. It shows the beginning of a great development. So is a shark an exhibit of an early development. How many improvements can you point out that other fishes have made on the shark model?

CHAPTER IV

THE DOG

THE dog was one of the earliest possessions of man, and, since it is largely this ownership of property that makes man different from the beasts, it must have helped him to get his early start toward becoming the dominant creature he is in the world. Man is the master mainly because of three things: he learned to use tools, he learned to make other animals help him with his chores, and he learned to cultivate plants.

Only one other animal, the ant, has had enough sense to do anything very much like man. The ant keeps its herds of cattle which it milks for honey. It plants and cultivates its mushroom beds. It goes on slave raids that it may have more labor with which to increase the output of its plants. It uses other ants as jugs, hangs them up in the pantry and stores honey in them, which is an approach to tool using. Had the ant been as big as man, it might have built itself a flivver, learned to handle a shotgun, and then it would have competed with him for the mastery of the world.

The dog was man's first experiment in training other animals to serve his purpose. Man was a wandering, shiftless nomad back in the early days of the race when he first, because of an awakening sympathy, tamed a wild dog and taught it to follow him from one camp to another. His undeveloped mind began to feel a responsibility for this creature which trotted at his heels. Responsibilities always start individuals to thinking, to

THE DOG WAS THE FIRST ANIMAL TAMED BY MAN AND TAUGHT TO SERVE
HIS PURPOSE

making plans for the future. The minds of members of these early tribes were stimulated by their having dogs.

Early man was a hunter. He soon found that the dog, itself a hunter, could help him in this work. There were certain things, such as speed in running, following a trail, in which the dog surpassed the man. He could use qualities of the dog in the chase. The dog had a better sense of smell than had man, and slept more lightly. It could warn man of the presence of other creatures that might injure him. It could be set to watch. Man had added to his power in the world.

It may well have been the taming of the dog that suggested to man that he might make similar friendships with other animals. The mild and pensive sheep, in the course of cycling centuries, gradually came to yield itself to the control of man. The donkey, the honey bee, the cow, the camel, as ages passed, came under his influence. They were creatures of much more use to him than was the dog, but the dog was the first, started it all. All brought to man powers that did not lie in his own frail body. Each was a responsibility that made him think; each helped him on toward ruling the world.

It is a matter of current belief that the dog is a descendant of the wolf, that wolves were tamed at some time in the distant past, that they have undergone certain changes while under the domination of man, and that the dog, as we know it today, is the result, is merely a trained wolf. There are those who differ.

Man, through the ages, has succeeded in domesticating not more than sixty of the thousands of species of wild animals. These have been species with certain instinctive traits. Practically all the domestic cats of the world, for instance, came from one wild species with a mild disposition. Individual elephants may readily be tamed, but will almost always refuse to bear offspring other than in the jungle. Parrots show intelligence by learning to talk and otherwise, but may not be bred in captivity. Many animals may be tamed, individuals brought to do the will of man, but not domesticated, brought to live generation after generation with man as if it were their natural condition.

It is probable that several species of wild dogs or wolves have been tamed at different times and places and have started generations of domestic dogs. The American Indian, for instance, had a dog when white men first came.   It is an odd fact that that dog has ceased to exist.   Early writers mentioned these dogs, but gave no details.   Before scientific men got around to studying them they had been displaced by the dog of the white man.   Nobody knows if they were a distinct breed or merely tamed coyotes or wolves. The dogs that have become domesticated, however, have undoubtedly been those wild creatures of their kind that have had mild, social and affectionate dispositions, that have been willing to yield themselves to the will of man. Most of them probably came from some one strain of a mild disposition which, like the Egyptian cat, came in contact with early man.

Pausing for a moment to give the dog its place in the scientist's scheme of things, we find that, together with the foxes and wolves, it forms a family in the order of carnivorous, or flesh-eating animals, of which the cat group is another outstanding family. Both cats and dogs belong to a larger grouping of animals known as a class, the mammals, or milk-giving animals, of which cattle, horses, and many other creatures are members. These, in turn, belong to a still larger group, the vertebrates—animals with backbones—including birds, fishes, and many others.

A partial outline of the Animal Kingdom would thus show the place of the dog:

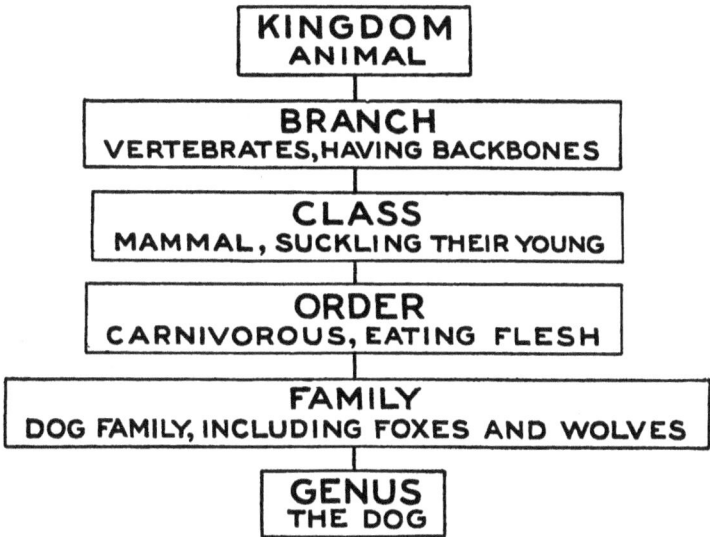

KINGDOM
ANIMAL

BRANCH
VERTEBRATES, HAVING BACKBONES

CLASS
MAMMAL, SUCKLING THEIR YOUNG

ORDER
CARNIVOROUS, EATING FLESH

FAMILY
DOG FAMILY, INCLUDING FOXES AND WOLVES

GENUS
THE DOG

The dogs of savage peoples are themselves usually very primitive and undeveloped creatures. Their masters have known nothing of the principles of selective breeding and such influences as appeared to improve the breed were accidental. The modern breeder knows how to make use of this principle of selective breeding. He knows that among white chickens, for instance, there may appear individuals with black feathers. If those with the most black feathers are selected and crossed, other chickens will appear with more black feathers. If those with the most black feathers are selected and crossed, generation after generation, a breed of black chickens will finally result. So by selection may certain qualities be bred into animals or plants.

Savage people, of course, knew nothing of this secret of stock breeding. It happened, however, that, among their dogs, conditions led to an accidental application of this principle. When savages roamed as nomadic tribes followed by their dogs, it often happened that famine came upon them. In emergencies of this sort they ate their dogs. The dog, for many thousands of years, probably served man as an emergency food supply.

Naturally, when these savages were driven to devour their pets, they ate first those that they loved least. These were the dogs that were least affectionate, least watchful, least useful. So it came to pass that the most affectionate and the most useful dogs remained to produce the next generation and make their outstanding qualities stronger in it. In this way and by accident did these savages tend to improve the quality of their dogs.

This law of selection worked even more effectively when man became an owner of flocks and learned to use dogs in helping to control them. The dog which was the most useful and had the greatest understanding about the flocks was the dog that was the longest spared and, therefore, the one that produced the most offspring.

The tending of flocks is almost the original occupation of man leading toward civilized life as we know it today. This process of selection among shepherd dogs has, therefore, been in operation for many thousands of years.

It is because of this that the shepherd dog of today is a creature of marvelously developed instinct that will look after sheep with almost manlike intelligence. It is because of this that a young shepherd dog which has never seen a flock, when brought into contact with one, will lose interest in everything else and devote its un-divided attention to it, will begin instinctively to tend it, to bring in the wanderers, to discipline them without injuring them, and to show a manlike affection for them, an affection which, by the way, these creatures of a different animal family seem to make no effort to pay back.

This process of accidental selection of the fittest as the parents for subsequent generations is also to be ob-served among the Eskimos where dogs have, for great periods of time, been used as draft animals in drawing sleds. The dog was undoubtedly the first animal ever put to this sort of work by man. The sled dogs of the North are specialized creatures with great strength and an instinct for their work which has been bred with them because their owners have eaten the bad dogs and kept the good ones to become fathers and mothers. The principle of the removal of the unfit is here applied also by the dogs themselves. A weak, sick, or injured dog, one not capable of defending itself, is likely to be set upon by the pack and torn to pieces. Thus do only the strong survive and thus is the strain steadily improved. Only the fighters survive, also, and thus these sled dogs, in their relations to each other, have become steadily more vicious. A side light on the method of primitive peoples that sometimes works out very well is here to

5

GREYHOUNDS CHASING A RABBIT

be seen. The sled dog must serve a certain purpose, must pull a load. When it is still a small puppy its master fits a harness to it and ties on a block of wood which the young dog must drag wherever it goes. Thus it grows up always pulling a load. Thus it learns its business.

Bulldogs were developed as cattle dogs, and in the centuries past have been trained to tend herds of cattle much as shepherd dogs tend flocks of sheep. The cattle of earlier times were less domestic than at present, were wild and fierce and unrestrained by fencing. Dogs of strength and determination were necessary to control them. Later bull baiting with dogs became a popular sport and for centuries dogs were bred with always greater strength and courage to take part in these entertainments. Thus did the passing centuries tend naturally to develop dogs of exaggerated traits along certain lines.

Take again the greyhound, most graceful and fleetest of all its kind. Lovers of sport, of the chase, through centuries bred these greyhounds for speed, to catch hares for their amusement. Fox hounds were bred for the acuteness of one sense, that of smell, that they might follow their game by their noses. Setters and pointers, with a rare talent for helping their masters find game, are most delicate examples of special qualities that may be bred into animals. Terriers have been given much skill in killing rats. Watchdogs specialize in repelling intruders. St. Bernards guard the passes of the Alps and rescue wanderers lost in the snow. Bloodhounds track escaped criminals.

FOUR TYPES OF DOGS

The dog has shown itself to be the one animal into which these unusual qualities could be bred most easily. Not only that, but its body has lent itself no less to modification by the breeder. The short-legged, long-bodied dachshund is a quite different creature from the Great Dane or the fluffy lap spaniel. The style of the beautiful collie contrasts oddly with that of the Missouri hound dog. The muzzle of the poodle seems grotesque beside that of the Russian wolfhound.

All these dogs with their many peculiar traits of body came from similar original stock. The body of the dog and its character have been particularly easy for man to influence. One reason for this, the scientists say, is the fact that the dog family is comparatively young in the world. Its nature has not become fixed.

Man also is a young species in the world. He, too, has shown himself capable of change, of development. It is different with such creatures as the cockroach, which has existed as it is for many millions of years, as is shown by remains of it found in coal deposits laid down ages ago. The nature of the cockroach could not be affected in a mere score of thousands of years.

Aside from being of a young race the dog was a carnivorous animal, accustomed in the native state to live by the game it caught. This in itself called for physical and mental activity. And it had by instinct that thing rare among animals—a nature capable of domestication. Thus it came, many thousands of years ago, before any other animal, to live with man, then just started on his climb to mastery. The two have lived together certainly for ten thousand years. They have come to understand each other as no other two animals. The dog has come to have almost human traits—such as no other animal.

The dog, for instance, establishes personal relations with the human beings about it. It has one master that it places above all others and its attitude toward that master is different from its attitude toward anybody else. To the immediate associates of its master it has a well-defined relationship. They are its friends, intimates, members of its family circle. But they are not the dominant influence in its life as is the master, to be worshipped and followed blindly.

Then there are the mere acquaintances of the master, people who come to the house, people for whom it has a friendly feeling, but toward whom it also feels a bit of formal reserve. One step further removed are the people whom it has never met, those that come and go incidentally—strangers. The dog's attitude toward these people is much like that of its master. It shows them a certain impersonal courtesy, but realizes that no relationships have been established. A dog would no more stare at a stranger than would a good-mannered human being.

Still further removed is the class of persons of whom the dog is suspicious, persons who are likely to be enemies of the master. It has no confidence in persons who move by stealth, persons who make their appearances by unaccustomed entrances, or ill-clad and suspicious-looking persons. Its judgments of such persons have become an instinct developed through thousands of generations in association with man.

There is no other creature that makes anything like the response to the presence of man that the dog does. When the master returns, note the joy of the dog. There is nothing else like it in the animal world. Where, also, is the creature that is so hurt by the master's rebuke and that can so perfectly register its humiliation? Where is a creature whose expression so changes at the sound of laughter, which can be so readily shamed, whose countenance so falls when addressed in a tone of melancholy? Where is the lower animal with the sense of ownership that will cause it to guard its master's coat?

To the student of dog life one sad fact seems to be developing. It begins to appear that the heyday of the dog at its best is past. The dog as a creature which serves a useful purpose is coming to be of less and less importance. Instead, it is growing more and more to be used merely as an ornament for the lawn or the benches of the Dog Show. As such the breeders are devoting themselves more to the development of oddities in the form of the dog than to an emphasis of its mental qualities or its practical usefulness. Breeders are finding better sales for grotesque specimens than for any other.

To the demand for the grotesque are brought the remarkable possibilities of modifying the body of the dog. The result is the development of such creatures as the toy dog, the pug dog, the glossy-haired spaniel, the coach dog, the nonsporting greyhound, even the lap greyhound, far removed from any possibility of usefulness and depending entirely upon size or form for their popularity.

When the care of cattle depended upon the sturdy qualities of the bulldog; when sheep were grazed in unfenced pastures; when the safety of the home depended more upon the vigilance of the watchdog; when the hunt had a dominant place in the everyday life of the people; sterling qualities of usefulness, strength, character, and intelligence came naturally to be bred for long periods into this animal which has always been classed as man's best friend. The conditions of modern life, however, are such that the dog tends to become merely a pet or a show creature and as such to lose much of the vigor of its past.

QUESTIONS

1. How did the man who first tamed a wild dog and had him trotting at his heels differ from his companions? Picture the sort of home in which this man probably lived. How do you think the dog helped this early man to get ahead?

2. Is the dog a tamed wolf? Can all animals be domesticated? What are some of the difficulties? What is the difference between being tamed and being domesticated?

3. To what order do dogs belong? what class? what branch of the animal kingdom?

4. What is meant by selective breeding? How did it come to work accidentally among dogs? What is meant by the law of the survival of the fittest? How did this work out among Eskimo dogs?

5. How was the bulldog developed? the greyhound? the St. Bernard? Are the poodle and the Russian wolfhound "sisters under their skins"?

6. Sharks, amphibians, turtles are very old groups of animals, have become set, and are hard to change. What are some of the young groups that change more easily?

7. How long have man and dog lived together? Compare the understanding shown by dogs that you know with that shown by cats, horses, rabbits. What conclusion do you reach?

8. If you have a dog or know one very well, make a list of that dog's human associates. Does he rank some one person first? Who comes second in his regard? Third? Fourth? Has he mere formal acquaintances that are on a different basis from strangers? Observe carefully this dog's attitude to different people.

9. Does this dog consider certain persons its enemies? See if you can find out why.

10. Experiment with your dog to see what effect mere tones of voice have on him, e. g., a scolding tone, a laughing tone, a melancholy tone.

11. Considering the recent development of peculiar varieties of dogs by the fanciers, what do you think the dog will be like two hundred years from now?

Chapter V

THE FRESH WATER MUSSEL

I T is a strange fact that we might not have pearl buttons to sew on our clothes if baby clams were not able to steal rides on fishes.

This single fact is a striking example of the manner in which the well-being of one living thing in this world is dependent upon those around it. It shows how each needs help and must have it to get along. Since each must be helped by others it would appear that each should be willing, in turn, to lend a hand to others, which teaches a lesson that might be applied by human beings.

This ride stealing of the button clam is one of those peculiar tricks of nature upon which individuals depend for their very existence. If the button clam fails to steal its ride it dies. As a matter of fact, the great majority of them do fail to get this ride and do, as a consequence, die. An occasional clam succeeds in getting aboard its peculiar sort of passenger train, riding as far as it likes and stepping off, like a settler who has gone west to a region of new possibilities. It prospers and carries on the race.

These button clams, or fresh water mussels, are found in most of the streams that flow into the Mississippi River.

If they are allowed to live until they are eight years old they develop shells, some of which are as big as plates. From such mussel shells many pearl buttons may be cut. As buttons go, pearl buttons are among the most attractive in all the trade. It is a fact, also, that buttons from these clam shells can be made quite cheaply as compared to those from other materials. Thus it has come about that pearl buttons are used today in great quantities all around the world and are more in demand than any others.

THE END OF THE MUSSEL—PEARL BUTTONS

The upper Mississippi Valley has an advantage over any other region in the world in the production of these buttons because of the great supply of raw material in the way of mussel shells. This fact was not appreciated, however, until 1890, when its discovery came about in an interesting way. A young immigrant had come to Iowa from Europe and was working as a farm hand. In the old country he had worked in a factory which cut buttons from horn. He had operated a machine which bored out these buttons. He examined the mussel shells along the Mississippi. It looked to him as if they would make good buttons and as if the abundance of them in this region offered a good business opportunity.

From this idea of an immigrant farm hand grew the button industry of Muscatine, Iowa, an industry which caused an activity along the Mississippi that for excitement suggested the gold rushes of the West of the early days, an activity which led to the establishment of many factories and which put many thousands of mussel fishermen into the streams, resulted in the gathering of fifty thousand tons of clam shells a year. These shells, converted into buttons, traveled in all directions from Muscatine until they reached most civilized households in all the world, and found places opposite most of the buttonholes on earth.

At first it was thought that the mussel shells of these Mississippi Valley streams would never give out. With the passing of two decades, however, it began to appear that the supply was running low. Under the circumstances, the Federal Government at Washington, as is its wont, began to inquire into the clam shell situation, and to study the possibility of taking action to keep up the supply. As usually happens, this study led to the discovery of some very interesting facts in the life story of one of those peculiar animals of the world which are not likely to be thought of as animals at all.

The Government studied the life history of the fresh water mussel. By knowing the history of the mussel, it figured that some way might be found to increase the number. The Government remembered that it had

found a way to decrease the fly nuisance by discovering
that flies bred in manure piles; to increase the fish supply
by hatching millions of them at the Government Stations.

It was found that the mussel is a prominent member
of one of those seven chief classes into which the
invertebrates, the animals without backbones, divide
themselves, thus:

```
                      [INVERTEBRATES]
                           |
[ARTHROPODS]  [MOLLUSKS]  [PROTOZOA]      [WORMS]
     [SPONGES]  [SEA ANEMONES]  [STAR FISHES]
```

The mussel is a mollusk, an animal wearing its skeleton
on the outside in the form of a protective shell, and its
soft, boneless, headless, almost formless, body on the
inside. This body is covered with a mantle, a sort of
skin which takes mineral matter, chiefly lime, from the
water, and from this builds this structure of shell.

When the investigators had followed the cycle in the
fresh water mussel's development, when they knew the
story from birth to button, they found it to divide itself
into stages somewhat as follows:

The clam first appeared as a tiny, microscopic animal,
already alive but undeveloped there in the pouches of
the mother clam. One mother would produce countless
numbers of these young. They would stay there in these
pouches until they had reached a certain stage of development when they could strike out for themselves. They
would then leave the mother never to see her again.

When these tiny clams left their mother's pouches they
were still so small that one would need a microscope to

see them. If they could be sufficiently enlarged, however, they would look like little clams with their shells
spread wide open, with most vicious teeth along the
edges of these shells, and with a little thread which

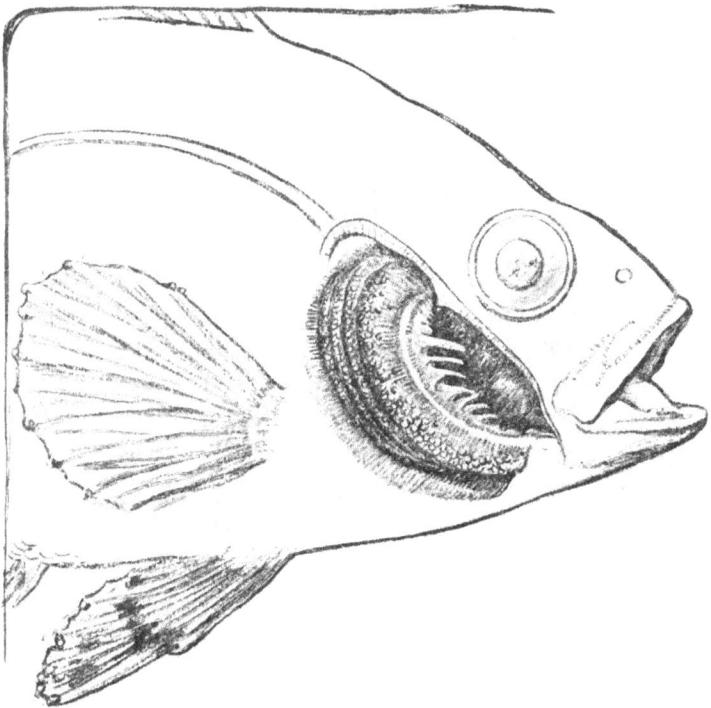

THE WAY A FISH CARRIES BABY MUSSELS ON ITS GILLS AND FINS

floated above them. In this form they would settle into
the silt at the bottom of the stream, send up this tiny
thread as a feeler, and lie there awaiting their opportunity to catch a ride upon some unsuspecting fish that

came idling past.   One in a thousand of them caught that ride.   The others died for lack of it, for one stage of their development is quite dependent upon fastening themselves for a time to a fish.

This lucky, tiny mussel which got its chance did so because the thread it sent up was drawn into the gills of the fish as it breathed and there caught hold.   It carried the baby clam with it and as soon as it got close to the fish's gills it clamped itself on with its vicious teeth. So small was it, however, that the fish scarcely realized its presence.   A single fish six inches long might breathe a thousand of these little clams into its gills, might have each of those thousand sink in its teeth, bury itself under its skin, and still be none the worse for the experience.   To be sure, the fish might be a bit uncomfortable, just as is a human being when he gets red bugs under his skin, but red bugs do no great harm.

This fish with its thousand passengers swims down the river, pursues its customary course of fly hunting among the cattails, or even travels to distant waters. Wherever it goes the baby mussels go with it, feed upon it, and develop there beneath its skin.   In the course of time they grow so strong that they no longer feel the need of being thus carried about.   They scramble out from their refuges, drop their comfortable sleeping quarters in the gills of the fish, and sink to a place in the bottom of the stream.   Each of the thousand emerges in its own sweet time and each finds its separate place upon the bed of the river.   Thus a fish serves a purpose of broadcasting these clams, of sowing over the bed of the stream what in time is to be harvested and converted into a button crop.

Those little mussels half bury themselves in the mud at the bottom, preferably in the part of the river which flows swiftly, for it is the flowing water that brings them their food. The mussel has a foot by means of which it can crawl, snail-like, on the bottom, but, as a matter of fact, it seldom moves. It lies there with its shell slightly open, the open part up. The ruffles of its mantle fluttering in the stream extract its food from the waters that flow through the mantle, and so the animal grows. There it lies until sometimes five or six years or possibly eight years have passed. Then the mussel fisherman, practicing his skilful art, appears and lures it to an act that brings about its own destruction.

Here is the way the fisherman catches the button clam. He goes out in his boat with a drag, a bar of gas pipe a dozen feet long, to which are attached many pieces of heavy fishing line three or four feet long. To this line are tied, every few inches, four-pronged hooks made of heavy wire. With this he drags the bottom of the stream where mussels are likely to be found. It is in the nature of the mussel, as the fisherman knows, to close its shell upon any object that drags between the two sides of it as it lies there with open mouth awaiting its food. So, when the iron hooks of this drag pass between the open valves of the shell they clamp shut upon them. They clamp so tight that the shell is pulled from its accustomed place in the sand and is dragged along the bottom. When enough shells have been caught the fisherman lifts the drag with the mussels still gripped to it into his boat and takes off his catch.

The Government, understanding this life cycle of

the fresh water mussel, has tried to work out some scheme for increasing the mussel crop of the upper Mississippi Valley streams. It has realized that this mussel crop is dependent upon the fish that broadcast the baby mussels. If it could bring it about that enough fishes went forth each year, each carrying its cargo of one thousand to two thousand parasite mussels, it could line all the streams with growing mussel shells. It found, first, that there must be fish or there could be no mussels, and, second, that these fish must have been given a chance to get baby mussels in their gills or they would not help the button crop.

The Government has a laboratory and experiment station at Fairport, Iowa, and there, for years, has been trying to work out the best methods for increasing the supply of fresh water mussels.

In the first place it gets in touch with the mussel fishermen who are bringing in their hauls, and gets from them the female mussels whose pouches are filled with the tiny young ones that are about ready to start out for themselves. It gets the young from these pouches and puts them in cans of water. It does not require many of the females to provide millions of the young ones.

The second necessity in planting a bigger button crop is to get the fish to be used as carriers. Even before this clam farming began, a widespread scheme had been worked out which resulted in the capture of great numbers of these fish. In the spring the Mississippi usually overflows. After the floods have receded it always happens that there are pools of water back from the channel, which grow smaller and smaller, until they finally dry up. These pools, naturally, contain fish. When the water is very low in them the time is ripe for catching these fish easily. Parties are sent out with nets and seines, up and down the streams, to scoop them up. Formerly they were merely brought to the main stream and released. Now they are held for use in the button campaign. Seining parties operate not only in these pools but at any point along the streams where the proper fish may be easily caught. The result is that the Government is able, at certain seasons, to obtain great numbers of fish to which the young mussels can attach themselves.

There are difficulties due to odd tastes on the part of these young mussels. The taste of different varieties of mussels as, for instance, the niggerhead clam, run to certain fish. The niggerhead will fasten its nippers into but one sort of fish in all the world—the river herring. It must have river herring or it will stay right there and die and the niggerhead race can go hang and cease to be. The pimple-back clam prefers catfish and goes riding on them in all the streams. The mucket, which is the best pearl clam, likes the game fishes, bass, crappie, sunfish.

It becomes plain, then, that if the little fellows are pimple-backs, it is catfish that must be provided as carriers, and so on, depending on the variety.

Thus, having the proper fish and the proper youthful mussels, it is only necessary to bring the two together under certain conditions. The mussels in the desired number are put in a tank in which favorable conditions are created. Then the captive fish are thrust into the tank and allowed to remain the necessary time to take on all the mussels they can handle. The experts know just how long. Then these fish are released in the streams. They go about their business and at the proper time will broadcast the developing clams.

In working out this scheme the Government has accomplished two ends. It has reclaimed many fish that otherwise would have died and it has increased the supply of the fresh water mussels. But, in spite of the Government's efforts, the clam beds have been overfished and it has become necessary to have closed seasons on first one section of the river and then another.

These fresh water mussels supply another product which is highly prized by man, another product which again comes into existence because another animal thrusts itself upon the clam much as its offspring imposes upon the fish. As the clam lies in the bottom of the stream, a parasite, a very tiny thing, may attack it. Just what happens is not well understood. Under certain circumstances, however, this attack of the parasite, or the presence of some other thorn in the flesh, hurts the clam, sets up an irritation within its body. That irritation may be between the mantle and the shell or it may be within the soft body of the clam.

The clam attempts to protect itself from this irritation. To do this it surrounds the thing that is hurting it with a layer of pearl, the material for making which it separates from the water. Thus does a tiny, smooth pebble of pearl start to develop within the clam. As time passes the clam deposits more and more pearl about this center and the size of it gradually grows. In the end it becomes that beautiful precious stone prized by jewelers,

6

the pearl of commerce, and may have a value ranging from a few dollars to thousands of dollars.

The hunt for these pearls in the bodies of clams has always been a part of the clam fisherman's business, has always offered him the chance of an unusual profit. The value of pearls recovered from the clams of the Mississippi Valley used to be, as a matter of fact, equal to about half the value of the clam shells themselves as they were sold to the button factories. When the return from clam shells was around $800,000 a year the return from pearls was likely to be about $400,000. Now, however, with the closer working of the beds, fewer clams live to be old enough to yield good quality pearls and their value is only about ten per cent of that of the shells.

The knowledge of the manner in which the mussel develops a pearl to relieve it of irritation from some object which hurts it, just as a cinder in the eye hurts, has led men to attempt to put into the bodies of clams certain objects that would hurt them and cause them to develop pearls when they otherwise would not have done so. This has been done with some success, particularly in Japan where there are large numbers of salt water clams which produce pearls. The Japanese have developed a method of so irritating these clams as to cause them to produce pearls. A profitable industry is based upon this method.

The oyster, which is an important food product, is a cousin to this fresh water mussel. Its shells are not, however, of such quality as to have any value as button material. Neither does it yield pearls of any value, although people everywhere who eat oysters continue to believe that they may some time find a fortune in their food. A pearl of value, be it known, was never found in the sort of oyster that is eaten.

Another interesting cousin of the fresh water mussel, or clam, is the scallop, found abundantly along the Atlantic coast, particularly about twenty miles out. Many people who eat scallops do not know and often wonder where they come from.

A big sea clam yields the food known as scallops. It is a very athletic clam, one which can perform feats impossible to most of its relatives. It can swim about quite freely in the water.

All the clams join their two shells together at the back by a hinge. The two valves, or shells, of the clam are tied together at the back by a strong piece of muscle by means of which the shells are opened and closed in feeding. The scallop is not satisfied with this limited use of its valves. Neither is it satisfied to creep along the bottom on its single, snail-like foot. It uses its valves as a means of traveling about. It opens them wide and snaps them together again with great force, and by repeating the process can go sailing through the water much as a quail goes sailing through the air. Doing just this develops a big knot of muscle where the two valves join. The scallop fishermen cut this muscle out, and it is the good white meat which one gets with tartar sauce at the restaurant.

QUESTIONS

1. All of us have worn pearl buttons all our lives. Of what are they made?

2. What is the chief source of supply for these mussels? Tell the story of the development of the pearl button industry.

3. To what "branch" of the animal kingdom does the mussel belong? -to what "class"? What are the peculiarities of mollusks? What mollusks do you know?

4. Trace the life story of the fresh water mussel. You have heard a good deal of the economy of Nature. Does it look to you as though Nature were economical when she makes it necessary that the mother clam should give birth to so many young ones in order that a few of them should grow up? Can you give other examples of the wastefulness of Nature?

5. Is the mussel an animal? How does it get its food? How old is it before it is grown?

6. Describe the method used by the clam fisherman to get his shells. Note that he depends on his knowledge of clam nature, the fact that the clams will snap shut on the prongs of his drag, to catch them.

7. What bit of knowledge of clam nature did the Government take advantage of when it wanted to increase the clam crop? Describe the plan for broadcasting baby clams.

8. What conclusion do you draw as to the possible usefulness of knowledge about even so obscure an animal as a clam?

9. Where do pearls come from? How are they made? How is man taking advantage of his knowledge of how pearls are made?

10. One often hears people talking of the chance of finding pearls of value while eating oysters. What does the author say of the probability of finding pearls in this way?

11. How do clams move? Which clams can "fly" through the water? What food for the dinner table do we get from this clam?

CHAPTER VI

THE GRIZZLY BEAR

HE grizzly bear is monarch of American beasts. During the centuries before the white man came he roamed about the western mountain wilds and wooded areas that made his favorite hunting ground, undaunted by any creature that he might meet. There he often found his nearest rival, the Rocky Mountain lion, which had pulled down a deer and was feeding upon its flesh. The great cat would promptly take flight at the approach of the masterful grizzly and the king was served.

Only the bravest of the Indians, then the only men on this continent, dared face the grizzly singly or in groups. The arrows from their bows often served to enrage this monarch and to make him dangerous to his attackers. A necklace of grizzly bear claws, therefore, was the most prized of manly decoration. Early white man, armed with the old muzzle-loading rifles capable of firing a single shot, found attacking the grizzly dangerous, for if that single shot did not find the vital spot, the wounded beast might turn upon the hunter and tear him to pieces.

The grizzly bear was master until man developed that

death carrying weapon, the repeating rifle. This dethroned the grizzly as king of the woods and changed him into a creature whose only chance to survive lay in his cunning in avoiding this man creature. Bold grizzlies that still refused to yield met death. The grizzlies that continued to live and breed their kind were those that fled into the fastnesses at the first taint of odor in the air, or at the crack of a dry twig, and by cunning avoided contact with man.

In the wilds of the Rockies there are still a few remnants of this lumbering, brownish, gray-tipped tribe, although they have entirely disappeared from great areas like the State of California, where they were once abundant. In parts of more desolate Alaska and Canada they still exist in considerable numbers. In such reserves as Yellowstone Park a few still live under protection. The one-time monarchs, however, are now fugitives. They are the noblest game of a continent and, therefore, are the creatures most desired by hunters armed with repeating rifles. So, year by year, their numbers are decreased and so is a great animal race being steadily pressed toward the time when it may no longer exist.

There is much difference of opinion as to whether or not the grizzly bear is an enemy to man which should be shot on sight. When early hunters met him on the trail and he failed to flee as did other creatures, those hunters, fearful of his power, immediately opened fire. The result was often a wounded and infuriated grizzly which charged its new-found enemy. At times also hunters met mother bears caring for their cubs. Such bears, fearing for the safety of their young, fought viciously. Again, a hunter might come upon a giant grizzly that had slain some game animal or had by force taken a carcass pulled down by a wolf pack and was sleeping upon it until it was devoured. It is instinctive with an animal to defend its food; and such a bear would fight all comers. Hunters did not understand the bear, giving no consideration to these special circumstances under which he should not have been disturbed. The result was often a tragedy. From such instances it was concluded that the grizzly would dispute the wilds with man. So man set out to destroy the grizzly.

The grizzly of today, according to those who have carefully studied him, offers little or no menace to the safety of man. Protected by a sense of smell that warns him of the approach of man, given eyes and ears of great sensitiveness, and having learned his lesson of the deadliness of this intruder, the bear vanishes when man approaches his haunts. Rarely does he attack. In Alaska, where he has learned his lesson less well, occasionally he still strikes. There is no proved case in the history of the West, however, of a bear having eaten the flesh of man. Hunters once found, for example, that in a storm a tree had fallen upon a prospector, his horse, and his burro, killing all three. Bears had devoured the horse and the burro, but the body of the man had remained untouched.

In Alaska in the last few years there have been several instances of men who have had the same peculiar experience with grizzlies. Each has been knocked down by a heavy blow from the paw of the great animal. In each case the bear has approached the body of the fallen man, has gone over it from head to foot, nipping it with his powerful teeth. The bear in this way has inflicted scores of flesh wounds upon these bodies, but in no case has he actually crushed and torn the victim with his teeth as he might so easily have done, and in a number of instances the wounded men have recovered. These men say that the bear, after nipping them, sat and watched them with great curiosity. If they moved, he came and nipped them again. If they remained still, he presently went away and did not return.

The bear is one of the most curious of animals and often acts in such a way as to give the impression that he has a sense of humor. One young bear, for example, will hide from another and peep forth inquiringly. A solitary bear crossing a snow field may be seen to stop suddenly, cock his head, and look curiously at his own shadow. Then he will pounce upon it and try to catch it. He will lie down, cover his eyes with his paw, and peep slyly forth. A bear, playing in the water, will stand a log on end, attempt to climb, it and splash noisily. He will climb to the top of a snow slide, start the loose snow, go tobogganing wildly down the mountain side, apparently better pleased if he can create a great commotion. Then he will examine curiously the track he has left. He will sit for hours watching the play of beavers. His attention is concentrated on anything that is new to him. He will creep from the woods, for example, and watch a fisherman casting in the stream. Many hunters have been surprised, on taking their back track in the snow, to learn that they were being followed for hours by bears. Bears have been caught peering curiously into men's camps.

There is a quaintness and humor about many of their actions. A young bear in Yellowstone Park, for example, found a camper's discarded bacon rind which was to him a fine prize. When a larger bear came his way the young one sat down upon the meat skin and began peering into a tree.

The family life of bears is odd. It is made up entirely of the mother and her cubs, as the father has no place in it. The cubs are born in the middle of winter, in the den in which the mother has stored herself away for the long sleep. They are very tiny, naked creatures, weighing but a pound at the time of their birth, and they do not get their eyes open for five weeks. Nature seems to have known that there will not be much food for these young bears and so has made them small that they may not need much. They may live for three or four months in these winter quarters with their mother, the only food supply being that which is stored away in her body, as she eats nothing during the winter.

When the spring comes the mother fares forth with her young, which are then not much larger than rabbits, and there begins a long romp of joyfulness which for them lasts about two years. Mother and cubs are constant companions. They are with her when she forages for food, when she explores the woods, when she swims in the streams. They watch her every move most curiously. They mimic her in a way that is quite comical. The mother, for instance, may be tearing a rotten log to pieces that she may get the grubs that are in it and of which she is very fond. The young bears watch so carefully that they get very much in the way. The mother is likely to cuff one of them roughly, knocking it sprawling. The young bear may pretend to be unconcerned, lying there where it has fallen, and may begin scratching busily as though for grubs. A faint scent of the man odor may here disturb the mother bear. She will stand upon her hind legs with her front feet folded across her breast and sniff the air inquiringly. Her cubs, assuming an expression of great wisdom, immediately do likewise.

REST FOR MOTHER BEAR—PLAY HOUR FOR THE CUBS

During all this time the youngsters never see their father. When winter comes again the mother establishes her den, taking the young bears with her. They are snowed in for another long winter and emerge to forage for another summer as a family group. At the end of this summer the cubs part with their mother, but are likely to remain together and to go into winter quarters together for another season.

The bear is a creature of the wild that is particularly easy to tame. Cubs caught when two or three months old very soon become as friendly as puppies. The bear is an animal that might easily have become domesticated. It has the friendly instinct that yields to domestication, an instinct limited to a few animals. The bear, however, is such a huge eater and is of such little use to man that it would be poor business to domesticate him. A herd of bears would eat man out of house and home.

During all this time the father bear stalks the solitudes alone, mating but once in two years for a brief period. Despite the humiliation that has come to him through man and his fire-spitting rifles, the grizzly is, in his way, still the monarch of the wilds. And among these old male grizzlies there is a peculiar sort of civilization, of recognized rights and reservations. Each bear, for instance, has his own domain. As he grows into bearhood, the establishing of this domain is his chief concern. He goes into the mountains and attempts to find a range which has not already been claimed. In this he decides to make his home. He goes about his claim and marks it out. His method of doing this is to rear himself upon the tree trunks and slash the bark with his claws and his teeth. Thus does he post his notices to other bears that this claim is taken. He scores the tree trunks with claws and teeth as high as he can reach and the height of these marks is the measure of his size. Another bear that may want to claim this range may come along, may measure himself against the score of the first bear, and, if he can mark the tree trunk higher, may remain and dispute the claim. The sturdy bear that towers above any other that comes that way is likely to be left unmolested.

The grizzly, once having marked out his domain, unless something happens to disturb him, spends the rest of his life in it. If any other grizzly intrudes, there is a battle for the mastery. To be sure, grizzly cubs may come into the range and remain without molestation until they reach maturity, for they are privileged characters. The solitary grizzlies, however, pay no attention to them, devote their lives to getting a living, to the solitary sports of their mountain tops, to an avoidance of the danger of encountering man intruders.

Grizzlies are ranked by the zoölogists as carnivorous animals, as eaters of flesh. They are not, however, eaters of flesh to the exclusion of all else, as are the cats. Their front teeth are the sort of teeth made for tearing which are peculiar to the flesh eaters, but some of their back teeth are flat, made for grinding, as are those of the animals that eat grains and grasses. The scientists would arrange them in the scale something like this:

CLASS
CARNIVOROUS

FAMILY
BEAR

GENUS
GRIZZLY

SPECIES
CALIFORNIA

The grizzlies of the Rocky Mountains eat great quantities of green foods, such as grasses, the tips of twigs, the bulbs of water lilies. Acorn and berry time in the forest is the time of feasting for the bears. It is upon berries and nuts that they get fat in the autumn in preparation for the long winter sleep.

When the bear family goes fishing, it furnishes good entertainment to any observer who may be sly enough to get a chance to see. Salmon streams of the Pacific coast

THE MONARCH OF AMERICAN BEASTS THE GRIZZLY BEAR

furnish the best setting for this angling of the bears. The young bears wait on the bank while the mother wades in. She puts her paws under the water and waits quietly until some fish comes near. Then she strikes a quick and mighty blow. More than likely one of her long claws will catch the fish and it will be flung among the cubs on the bank. If they squabble over it the mother will come ashore, cuff them into good behavior, and return to her fishing. After all the cubs are served, the mother catches a fish for herself.

It is an amusing sight to see a huge bear in a meadow catching grasshoppers on the wing, a feat of which he is capable and to which he devotes many hours on a summer day, for he is very fond of insects. Armed as he is with huge claws, the bear can dig as can few other animals and likes nothing better than to find the holes of chipmunks, gophers, and even field mice, that he may dig out and devour these rodents. Huge as he is, the bear is a great mouser.

He is little given to hunting big game. There was a time when he followed the buffalo herds and pulled down an occasional bull, but he probably fed more often upon animals which had met death by accident or otherwise. The bear is much given to taking the game that other creatures have killed or devouring that which has met death by accident. There is an occasional cattle-killing bear, as there is an occasional sheep-killing dog or chicken-killing cat, but he is an exception. In the West cattlemen have had much trouble with wise old outlaw bears with dens high in the mountains from which they have raided herds for years and defied all hunters. Usually rewards that tempt some skilled hunter to devote sufficient time and effort to find and kill the renegade are offered.

In the Rocky Mountains the bears at the approach of

winter usually dig for themselves caves in the mountain side into the mouths of which the snow drifts, thus sealing them in for the season. The polar bears of the far North take less trouble. They find a place under a shelf of rock or ice that may offer some degree of protection, or they scoop out a bit of depression, possibly merely in a snowdrift, lie down at the coming of a snowstorm, and let nature spread thick, white layers of blanket above them. In such a den a mother may lie for six months with her cubs, the flame of life burning low, her only connection with the outside world being a tiny vent hole through which the steam of the inside warmth and the strong bear odor may occasionally furnish a ·clue to the bear hunter.

This practice of hibernating, or becoming inactive for the winter, is followed by a good many animals. The principal idea is to get through a long period of food shortage. The woodchuck, for instance, is the best sleeper of them all. In the northern United States he gets very fat in the autumn on red clover, goes to sleep in October, and does not wake up until April. During this time his temperature is very low, his heart barely beats, he scarcely breathes. Because of this inaction his body uses up his stored fat very slowly. He comes out in April thin, but in good health. The ground squirrel, the chipmunk, and the raccoon thus hibernate, as does the giant, the bear.

The polar bears are cousins to the grizzlies, but, living constantly in the far North, they have come to be white that they may not be so easily seen by the wild life upon which they prey. They, too, are monarch of their domains and creatures of great prowess. They live among the ice floes and feed upon seals and fishes. They are as much at home in the water as on land and have been seen swimming freely fifty miles from shore.

Where the Aleutian chain reaches out toward Japan lies Kodiak Island. Upon that island dwells a bear that takes its name from it and is known as the Kodiak bear. It much resembles the grizzly, but is larger, a huge creature often weighing a ton, the biggest flesh-eating animal in all the world, living chiefly by fishing. If you should go to Kodiak Island the natives would tell you that not so long ago these bears were very numerous. They are still plentiful, for that matter, but so many hunters have come this way for the thrill of killing one of these master carnivores that their numbers are rapidly dwindling.

The brown bears, the grizzly bears, the black bears, and the polar bears are the four members of the bear family that are natives of North America. Of the four, the black bears, with their smooth, glossy coats, were the most widely distributed. In the beginning they were to be found wherever there were wooded areas, except in the far South. The increase in the numbers of men has driven them back, but they are still likely to be found wherever there are wild and desolate areas. They are very shy and flee in almost ridiculous haste whenever their sharp noses tell them of the approach of man. But for their wisdom in this respect they would probably have ceased to exist long ago.

The black bear is about five feet long from nose to tip of tail and likely to weigh from three hundred to six hun-

dred pounds. The grizzly is twice as large. Skins have
been taken that were ten feet long, and bears have been
weighed that were somewhat over a thousand pounds.
California grizzlies often weigh fifteen hundred pounds.
Ordinarily, however, a good-sized female grizzly will weigh
seven hundred pounds and the same sort of male will tip
the scales at nine hundred. The polar bear is about the
same size as the grizzly.

QUESTIONS

1. How does the grizzly bear rank among American beasts? Compare
   the numbers of them now in the forests with the number that
   used to exist. What influences have worked against the bears?
2. Is the grizzly bear an enemy of man? Give reasons for your answer.
   Do you think the grizzlies should be protected or wiped out?
   Give your reasons.
3. What makes you think that a bear has a sense of fun like that of
   human beings? Tell of any acts of bears that you have observed
   that seemed to show that they appreciated fun.
4. Where does the career of the young bear begin? Describe its early
   life. Which parent is its teacher? How is the young bear dis-
   ciplined?
5. Describe the manner of life of that solitary hermit, the male grizzly.
   How does he establish and keep his claim?
6. To what class of vertebrates do the bears belong? Are they more
   nearly related to cows or to cats? Give reasons for your answer.
7. Upon what do bears chiefly feed? They are classed with the flesh-
   eating animals. What do they eat besides meat? What does
   "herbivorous" mean?
8. What is meant when one says that an animal hibernates? Do you
   know of any animals that do so? insects? Describe the hiberna-
   tion of the bears.
9. Are there different species of bears? What kinds have you seen?
   What kind does the organ man lead about?
10. What is the largest of the bears? What makes the coat of the
    polar bear white? Do you think this polar bear is properly a
    land or a water animal?

7

11. What is the history, in America, of the black bear, so familiar to the early settlers?
12. Keeping in mind all you have learned about the bear, where would you place it in the scale of animal intelligence? What animals, if any, would you put above it in intelligence?

CHAPTER VII

THE RATTLESNAKE

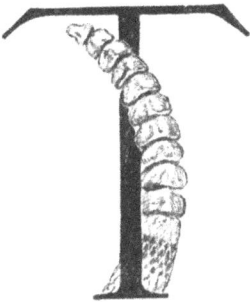

THE Frenchman was a scientist and he had witnessed the Hopi snake dance in Arizona, in which the Indians, wise in the knack of handling mad rattlers, had taken them in their hands, and even in their mouths, and had not been bitten.

The foreigner had watched and had seen how this thing was done. He, a white man, a scientist, could perform any feat that an untaught Indian could. Here now was a western diamond-backed rattlesnake from farther south, a snake nearly six feet long, of the royal family of the poisonous reptiles, which he would pick up even as had the Hopi. The Indians, he had observed, had teased their snakes until they struck, thus straightening out their bodies. Aside from a certain side stroke they had little power of action when they were not coiled.

The Frenchman waited until the diamond-back had struck. When it was laid out there on the ground like a rope, he grabbed it by the neck. This was as the Indians had done it. A snake so held could not reach the hand of its captor to bite it.

But the reptile writhed in anger, and buzzed loudly

with its rattles.   It looped its body around the French-man's arm.

And then, had one watched closely enough, he would have seen a very peculiar thing begin to take place. He would have seen that the part of the snake between the Frenchman's hand and its head was growing longer.

A snake can draw up the muscles in a given part of its body and make that part big.   Then it can relax that

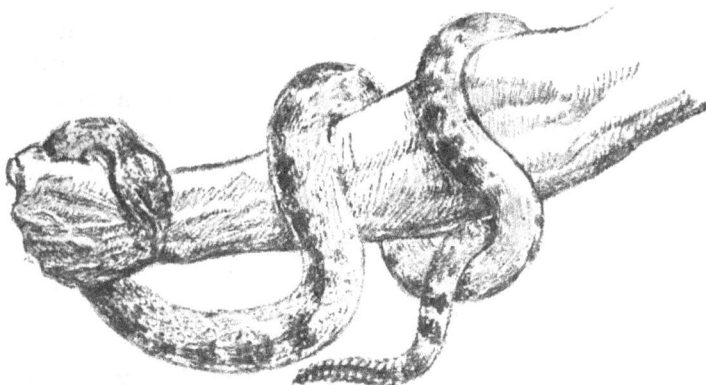

THE REPTILE LOOPED ITS BODY ABOUT THE FRENCHMAN'S ARM

part of the body and make it small.   It can draw in its muscle from behind an obstruction and push it in front of it. In this way a snake can crawl through a hole that is smaller than its body.   If one grips the rattler close to its head, however, one does not have hold of the muscular part of its body and it cannot resort to this device.   But this snake was coiled around its captor's arm and was steadily pressing its body forward.

So, in a little while, this diamond-back had several

inches of its body through the Frenchman's hand. When it had enough to allow it to bend its neck, it did so and sank its fangs into his arm. In a few hours the man was dead.

This diamond-backed rattlesnake stands at the head of a peculiar tribe. It finds its greatest development in the southeastern part of the United States, where it sometimes grows to be eight feet long and is at the same time heavy bodied. It is the weightiest poisonous snake in all the world. The southwestern variety, the kind of snake which bit the Frenchman, rarely exceeds six feet in length. Either of these cousins, however, is deadlier than any other rattlesnake because, being larger, they inject more poison.

In all there are some twenty species of rattlesnakes in North America, and there is hardly a region between Canada and the Gulf in which they do not live. Among the best known of them is the timber rattlesnake, sometimes called the banded rattlesnake, which is to be found along the Atlantic seaboard from New England to the Gulf. The odd thing about this snake is that the males may be black or almost so, while the females are a brilliant yellow. The prairie rattlesnake is a widely distributed species half the size of the diamond-back. The Pacific rattlesnake is a similar species, while the little sidewinder of the southwest is yet another—a pale yellowish or pinkish creature out of the deserts. Then at the other end of the range is the pigmy rattlesnake, so tiny that its alarm can be heard but a few paces, and it produces such a small dosage of poison that its bite is little more dangerous than the sting of a yellow jacket.

All these rattlers are snakes peculiar to North America. They exist at no other place in the world. The one thing that sets them aside from all other snakes is this rattle, this alarm clock on the end of their tails, with which they give warning of their presence. It develops there at the end of the tail, being made of the same material as is the top skin of the snake, of the same material, in fact, as our fingernails. Dry and bound together in loose joints, the parts of which will rattle against each other, this addition to the tail of this snake becomes an odd and freakish thing, such as no other animal possesses.

In rattlesnake country everywhere one will hear the story advanced that the age of a snake may be told by the number of rattles it has, and that each rattle marks its passing through a year of time. This theory of a relation between rattles and age was generally accepted until snakes were kept under observation for long periods. Then young rattlesnakes were born in captivity and grew to be big rattlesnakes under the eye of their keepers. It was found that a snake three or four years old might have a dozen rattles, and it was found that a new rattle was added every time the growing snake shed its skin. This skin shedding took place about once in three months if the weather was warm, but might be retarded if it was cold. A fairly safe way to estimate the age, based on rattles, it would seem, would be to figure that the snake produced an average of three of them a year, at least during its youth.

Scientists have long debated the purpose for which the snake used its rattle. Some have argued that it used it in terrorizing the creatures that it wanted to capture for food, but they have not even proved that it employed its rattle in hunting. Others have argued that the rattle was used by the snake as a call to its mate and played its principal rôle in its love-making. This is true of the katydid and the harvest fly, noisiest of insects, which are silent except at mating time. It is in the mating season that the frogs and the birds are noisiest.

More generally accepted, however, is the theory that this snake uses its rattle for advertising purposes. It is a deadly viper and most of the other animals of the wild have come to know this and instinctively to fear it. The rattlesnake is a clumsy animal and might be trod upon and injured by its larger associates if it were not feared and if it did not have some way of making its presence known. If one is riding on horseback along a path, for instance, and a rattlesnake sounds its alarm close beside that path, the horse is likely to leap frantically away from the sound. Thus an example is given of the manner in which the snake, by giving warning, has protected itself from being trampled.

To be sure, with the advent of man, this alarm of the rattlesnake is likely to cause its death, for man, recognizing it as an enemy and locating it by its rattle, will kill it. When nature developed the rattle for the snake, however, this was not a man-controlled world and this modern danger from undue advertising did not exist.

The dramatic and tragic thing about rattlesnakes, of course, is the fact that they bite and that their bites sometimes result in death. The manner in which they bite, the reasons why these bites result in death, and the way in which death can be prevented, is the vital part of the whole rattlesnake story.

The rattlesnake is the possessor of two poison fangs, one at the point of each upper jaw. They are more than fangs, they are hypodermic needles, teeth with tiny channels through them which come to the surface as they approach the tip and are continued in a groove that goes all the way to the tip. Back of the fang root is the poison sac in which the poison is secreted and held in reserve. As the fang is driven into the victim, pressure is brought to bear on the poison sac and this forces the deadly, clear, liquid poison through the hollow and groove of the tooth and into the wound. The diamond-back can inject half a teaspoonful of it and the smaller snakes in proportion to their size. The chance of recovery depends on the size of the dose.

A casual observation of the mouth of a rattlesnake would never reveal these fangs. They fold backward into the roof of the mouth and are only brought out when the snake is angry. When it strikes, the fangs come out like two needle-sharp daggers. They are driven home with a bayonet-like thrust straight ahead. Rattlesnakes do not bite. They stab. At the end of the stroke they squirt their poison. If the blow falls short they may spray the poison for two or three feet. If it goes home the poison enters the flesh of the victim.

The rattlesnake is very clumsy and travels quite deliberately. It cannot run to escape its enemies or in pursuit of its game. It is a creature of quietude that lies in wait.

The greatest activity of which it is capable is that of

coiling and striking. It winds the back half of its body in a coil which forms a base upon which the front end may work and in the middle of which the tip of the tail, when it rattles, gives warning. The front end rests upon this base, watchfully and threateningly, awaiting any intruder who might harm it. It rarely attacks, merely defends itself. If an enemy thus warned comes within the circle the snake strikes.

Wild animals, however, with the exception of those that prey upon the rattler (for even this deadly snake is eaten by some of them), keep out of the danger circle. This is easy, for even a coiled rattlesnake can reach no more than a third or at most a half of its length when it strikes. Man knows the limits of the danger zone and does not trespass except by accident. He is rarely bitten by a rattlesnake unless he steps upon it or puts his hand upon it by accident. Knowing of the presence of a rattlesnake, man kills it with ease and at no danger to himself, by the simple method of taking a stick and beating it to death. Stories of battles between rattlesnakes and human beings are pure fancy. When a human being sees a rattlesnake there is no difficulty in killing it.

It is not often that human beings die of rattlesnake bites. The bites of the larger snakes are dangerous, but not necessarily fatal. The bites of middle-sized snakes sometimes, but not often, lead to death. The bites of smaller snakes cause inflammation and a good deal of suffering, but are almost never fatal. Estimates made by scientists who have devoted their lives to a study of reptiles, lead them to the conclusion that the bites of these snakes are not often fatal. Some place the death rate as low as two per cent of those bitten, while others place it as high as ten per cent.

The danger is sufficiently great, however, so that every individual should know how to render first aid and should be ready to act with energy in case of rattlesnake bite. He should realize that a poison has been injected into the flesh and that the first object is to keep that poison from getting into the system. A snake bite is likely to be on an arm or a leg, and the first act, therefore, should be to place a ligature, or a band, about the limb above the bite so tightly as to stop the blood in that limb from running back to the body. A ligature may be made by tying a handkerchief or a cord about the limb and twisting it with a stick. Thus the poison is held in the limb.

The next act should be to make the wound bleed freely, that the poison may be washed out of it. To do this the wound should be slashed vigorously with a knife; it should be cut across first in one direction and then in the other. No attention should be paid to the mere pain of a flesh wound, since this may be a matter of life or death. After the wound is slashed it should be squeezed and pressed and otherwise caused to bleed freely.

The mouth of the victim, if he can reach the wound, should be applied to it and it should be sucked vigorously. If he cannot reach the wound with his own mouth his companions, if he has any, should suck it. It is an odd thing that this snake poison is not dangerous if swallowed. A man would be uninjured if he should drink all the poison in the glands of a diamond-back. If the skin is broken at any place within the mouth of the person who sucks a wound, however, the poison entering the circulation at that point may cause inflammation, but there will not be enough of it to become dangerous.

The ligature should be loosened in half an hour to relieve the pressure of blood in the limb and then put back again for a similar period.

Whoever is going on a trip where rattlesnakes may be found will do well to take with him a few crystals of a very simple medicine which will do more to kill their poison than any other. This medicine is potassium permanganate. It comes in the form of crystals and enough of it to treat a rattlesnake bite could be carried in a paper in the back of one's watch. These crystals should be rubbed into the wound at the earliest moment. They are dissolved in the blood and counteract all the snake poison with which they come in contact. After the wound is slashed they should be again rubbed into it. Potassium permanganate is a very ordinary chemical and can be procured at any drugstore.

One of the greatest mistakes that could be made in the treatment of snake bite is the program, so often followed, of giving the victim large quantities of whisky. Whisky thus administered has led to the death of many more people suffering from snake bite than it has saved. It has no effect on the poison, does not counteract it, has no place in its treatment. It but complicates a situation that is already difficult. It but adds to the burdens of the system, and increases blood pressure and tendency to coma that the victim has to bear.

There is never any question of the identity of the rattlesnake. Anybody can recognize it. This is one good turn which Nature did for man when she put the rattle on the end of the tail of this poisonous reptile. The most widely distributed of the poisonous snakes in the United States are rattlesnakes. Thus is the general problem of identifying this far-flung reptile solved.

There are two other snakes, however, cousins to the rattlers, that do not have this noise-making machine. They are the water moccasins and the copperheads. These two are much more closely related to each other than to the rattlesnake. The copperhead is, in fact, often called the upland moccasin.

The moccasin is primarily a water snake which inhabits swampy regions along the Gulf Coast and lower Mississippi. It is familiarly known as the cotton-mouth because of the white that is shown when it opens its jaws. The copperhead is very much the same sort of snake, except that it lives in the hills, preferring a rocky country. Both have the general characteristics of their relative, the rattlesnake, except that they are more active. Both strike much as does the rattlesnake. Both have the same sort of fangs and poison glands and inject the same sort of poison. This poison is milder, however, in the case of these snakes than in the case of the rattlesnakes. Death from a copperhead bite is not often reported. Bites from cotton-mouths are not frequent because they live chiefly in uninhabited regions. They are larger snakes than the copperheads, however, and therefore inject more poison. The copperheads may be three feet long, but are usually smaller, while the moccasins may reach a length of four feet.

There is one prominent mark of identification on all

these related poisonous snakes that is possessed by no other snake in America. That mark is stamped on the face of each as if it were a brand by which it could be identified. It takes the form of a deep pit on each side of the head, halfway between the eye and the nostril and a little below the line between these two. This pit is a deep depression half as big as the eye, very easily seen, and unmistakable. It is the mark of poison. The fact should be impressed on the mind of every individual that this

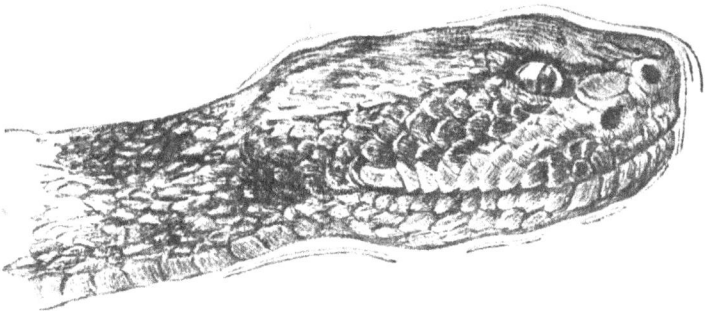

HEAD OF THE PACIFIC RATTLESNAKE SHOWING "PIT" OR INDENTATION OVER THE POISON FANGS

pit on the nose of a snake is a danger sign. Every snake which possesses this pit possesses fangs and poison glands back of them. Every snake with this pit is an enemy to man and should be killed on sight.

On the other hand, it is well to bear in mind that all North American snakes without this pit, with one single exception, are harmless. There are one hundred twenty-four species of snakes in the United States, of which seventeen are poisonous, and one hundred seven harmless.

These one hundred seven species of snakes are not only harmless but many of them are beneficial to man. They are mostly his friends. They should not be killed, but protected. Only the pitted snakes and one or two others should be killed. Snakes in general should be spared. They are among the greatest enemies of the rodents, particularly rats and mice, and these, aside from certain insects, are man's most destructive animal foes.

The familiar king snake, widely distributed in the United States, is one of man's prize friends in this work of killing pests. The king snake not only eats rats, but will feed upon as dangerous a creature as the rattlesnake itself. There have been eyewitnesses to attacks by king snakes upon rattlesnakes. Scientific observers have seen the former, more active than its poisonous rival, circle round and round the rattlesnake as it coils to give battle. The king snake awaits its opportunity and seizes the rattlesnake by the back of its neck, coils about its victim, crushes, and kills it. Then, as is the way of snakes, it lathers its victim with saliva to make it slippery, and swallows it whole, bones, rattles, poison glands, and all. After that it goes away to sleep for a month while the rattlesnake is being digested.

The blacksnake is also a great forager and is known upon occasion to attack rattlers. The beneficial influence of the blacksnake is questioned, however, because it devours many toads which help keep down the insect hordes; and because it climbs to birds' nests in trees, and swallows the young ones.

An unexpected enemy of the rattlesnake is found in the

hog which seems to be immune from its poison and which promptly devours it on sight.

There is one poisonous snake in the United States that does not belong to the pit vipers. It does not have that hole between its eye and nostril as a danger warning. This is the coral snake, or Harlequin snake, of the Gulf Coast. There is another obscure, related snake in southern Arizona, but it hardly deserves mention. This Harlequin is a pretty little crimson, black, and yellow banded snake which looks very much like certain innocent snakes of the same region, particularly the king snakes. It is hard to distinguish a coral snake from some of these harmless snakes without examining its teeth. The poisonous snake has a single fang on each upper jaw, while those that are not poisonous have a number of smaller teeth, but no fang.

This little coral snake, oddly, is a cousin of the terrible cobra of India which causes the death of twenty thousand people a year—there in the bare-legged country where snakes may bite so handily. It is the only North American relative of these snakes of the East. It is strange that the poison which it injects should be exactly the same poison as that of the cobra so far away, and quite different from that of the rattler. The pitted vipers manufacture poison by a recipe that is different from that used by the cobra tribe. Wherever members of either breed are found anywhere in the world, however, they produce the poison peculiar to their kind. Rattlesnake poison is ninety per cent blood poison and ten per cent nerve poison. It causes great local inflammation. The poison of the cobra breed is ninety per cent nerve poison and only ten per cent blood poison. Thus it causes little local inflammation, but it does cause a strong tendency toward a paralysis, particularly of heart action. The bite of a coral snake should get the same first aid treatment as that of a rattlesnake. Potassium permanganate is the proper antidote for its poison. The immediate attention of a doctor and skilled stimulation of the heart is more necessary, however, since the little coral snake is more deadly than the rattler. Fortunately it is present only in the extreme South and even there is seldom seen because it lives largely underground.

Snakes are reptiles, and reptiles are one of the five great "classes" into which the animals with backbones are divided—the mammal, bird, reptile, fish, and frog classes. Then, as the scientists classify living creatures, we may trace these snakes down until we get to the diamond-back itself. The descending stairway would look something like this:

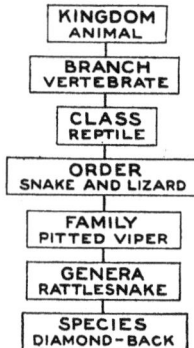

```
        ┌──────────────────┐
        │     KINGDOM      │
        │     ANIMAL       │
        └──────────────────┘
         ┌──────────────────┐
         │     BRANCH       │
         │   VERTEBRATE     │
         └──────────────────┘
          ┌──────────────────┐
          │      CLASS       │
          │     REPTILE      │
          └──────────────────┘
      ┌──────────────────────┐
      │       ORDER          │
      │  SNAKE AND LIZARD    │
      └──────────────────────┘
          ┌──────────────────┐
          │     FAMILY       │
          │  PITTED VIPER    │
          └──────────────────┘
          ┌──────────────────┐
          │     GENERA       │
          │  RATTLESNAKE     │
          └──────────────────┘
          ┌──────────────────┐
          │     SPECIES      │
          │  DIAMOND-BACK    │
          └──────────────────┘
```

Lizards, like snakes, are reptiles. The two families are related. The snake got the habit of wiggling along without its legs, and these, not being used, wasted away. On the sides of many snakes, particularly the pythons, are to be seen knobs, the remains of what were once legs.

The reptiles are a waning race. There was an "age of reptiles" when the world was a jungle overgrown with dense vegetation which was inhabited by huge reptiles, some of them bigger than elephants. Then conditions changed. The continents were elevated, the climate changed, a new class of animals, a fitter class, the mammals, developed, and has been steadily driving out the reptiles, killing them off. The mammals are a much later development of animals than the reptiles; they are later models. They are superior, just as later models of an automobile are better than the early models. The reptiles have been dying out gradually for millions of years. Their place in a modern world is not very secure. Only remnants of them are left. The newer race is displacing them. Now the mammal is king. Long live the King!

QUESTIONS

1. How may one hold a poisonous snake in his hand without being bitten?
2. What first-hand information have you on rattlesnakes? Have you ever seen one? killed one? If so, what kind of rattlesnake do you think it was?
3. What is the habitat of the rattlesnake? What sets it aside as being different from other snakes?
4. Can you tell the age of a rattlesnake by its rattles? What do you think is the real purpose of the rattle?
5. Describe the manner in which the rattlesnake drives home its fangs. What is it that makes them deadly? Describe the working of their hypodermic needles.
6. Does the bite of the rattlesnake mean death? What percentage of those bitten die?

8

7. What is the very first thing to do in case of rattlesnake bite? After the blood is stopped from going back into the body, what is next? What should be done to the wound after the bleeding? Why is there no danger of being poisoned from sucking a rattlesnake bite?

8. If you know in advance that you are going into a rattlesnake country, what is the medicine you should carry? Is whisky a help when bitten?

9. The rattlesnake is a pit viper. It has two cousins that are also pit vipers. What are they? Where is each found?

10. When you see a snake how can you tell whether or not it is a pit viper? When you see the pit what conclusion do you draw?

11. There are one hundred twenty-four species of snakes in the United States. How many of them are harmless? Why should they not be killed?

12. What cousin of the deadly cobra is found in the United States?

13. Is the rattlesnake an animal? Is it a vertebrate? Is it a mammal? To what "class" of the animal kingdom does it belong?

14. What other animals are the nearest relations of the snakes? Are reptiles an old or a young race? Are they gaining or losing strength?

CHAPTER VIII

THE HORSE

HERE was a time on the face of the earth when the ancestors of the horses of today lived in trees. This was during that period, æons ago, when the reptiles, huge creatures like the dinosaurs of our museums, ruled the world. It was the age when vegetation grew rank, an age from which has survived but one plant of today, the big tree of California, and from which has come down unchanged but a single creature of the animal world, the venerable cockroach.

These huge reptiles lived upon the ground and were so powerful that other animals that would survive had to take to the tree tops. This was the time when the reptile, itself a vertebrate, a creature with a backbone, began its age-long conflict with the mammals, also vertebrates, for dominance. Those mammals, thus driven to climbing trees, were the ancestors of the mammals of today, such as cats, cows, bears, hippopotami, man himself. They had five toes, given to grasping, on each of their four feet.   There in the tree tops they kept alive the germ of mammal life which was destined some day to rule the world.

As time passed the reptile race waned, and the mammals were able to come out of the tree tops and exist upon the ground.   There they had less use for feet adapted to purposes of grasping and more use for those that were fitter for walking.   As the centuries passed and these animals walked more and climbed less, the structure of their bodies changed.

In the rocks of past ages, in coal beds, in asphalt deposits, are to be found the remains of the animals that lived in given ages.   By examining these remains the scientists have been able to trace the gradual change in this creature which became the horse from the time when it lived in trees.

Early among these remains is a little animal eleven inches high with four toes on each of its feet, shown by its form to be closely akin to that ancestor which had swung in the trees.   Then later appeared this same creature a little further developed, as big as a shepherd dog, with three toes all of which touched the ground when it walked.   A yet later specimen still has these three toes, but the one in the middle has grown much larger and the other two were unused and tending to disappear.   Finally there is the horse of today with but a single toe, yet carrying within its feet bones which indicate the one-time presence of now discarded toes.   One of the most remarkable examples of reversion to type and present-day proof of this origin of the horse is the fact that occasionally into the modern world there is born a horse

which has the three toes of its ancestors, two of them
small and useless, yet none the less distinct toes.

It is interesting to note that when the ancestor of the
horse came out of the tree tops and began to walk upon
the ground, there also appeared a creature very much
like it which, in its development, took a slightly different
turn. Instead of discarding all of its toes but one, it retained two toes of equal strength. From it has grown a
different family in the animal world, that split-hoofed
family of which the cow, the deer, the pig are the best
known examples.

These hoofed creatures are all herbiverous animals,
eaters of grass. The flesh eaters lived in the trees and have
shown less tendency to leave them. The cat family is
still to a degree a tree dweller, but the dog family has come
to spend its life entirely upon the ground. The bear
family, on the contrary, is just now in an interesting
stage of change—of transition—still given to climbing
trees, though chiefly an earth dweller, still grotesquely
awkward in its movements on the ground because its
feet and general structure have not yet developed into
efficiency for that sort of life. Man, also, descending
from these tree dwellers, has retained many of their characteristics, chief of which is the structure of his hand.

Of all the animals in the world, the horse has developed
the most efficient feet for rapid and enduring use for travel
on the ground. As the ages passed the nail upon what
was originally its middle toe has developed into one solid,
strong, weight-bearing hoof, which will support its load
to much greater advantage than any other foot of any
living creature.

And, as the horse developed this master foot among the animals, it also came to have a form of greater symmetry and refinement. The body of the cow, for instance, is a crude and rough thing compared with that of the horse. The cow, consequently, is not capable of the speed or endurance of the horse. Its feet, primarily, will not stand the hardships of the road.

Scientists find the place for the horse in the scale of animals something as follows, its chief difference from the dog and the cat lying in the fact that it eats vegetable food and not flesh:

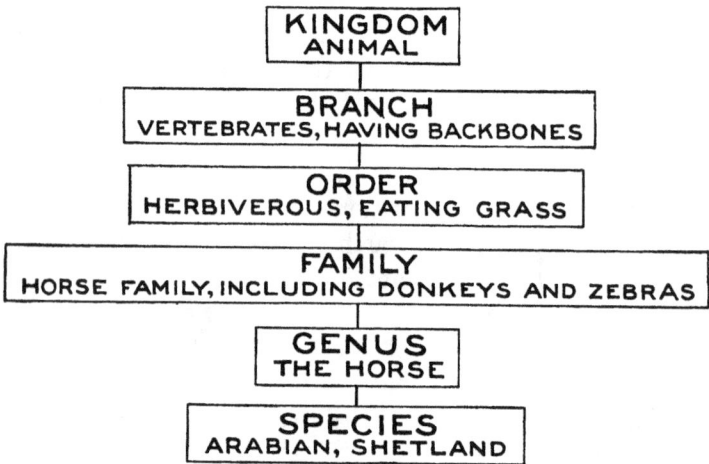

KINGDOM
ANIMAL

BRANCH
VERTEBRATES, HAVING BACKBONES

ORDER
HERBIVEROUS, EATING GRASS

FAMILY
HORSE FAMILY, INCLUDING DONKEYS AND ZEBRAS

GENUS
THE HORSE

SPECIES
ARABIAN, SHETLAND

In the case of no other animal is there so clear a record of evolution as in that of the horse. Oddly, the greater part of that record is written in the geological history of North America. The great plains of the United States furnished the stage for the development of the horse.

Here it grew into a stature and prowess quite equal to that of the wild horses of Asia from which the domesticated animals of today mostly descend.

At this stage of the horse's history, however, there occurred an incident which is as yet not explained and which undoubtedly changed the course of the development of the world. The horse disappeared from America where it had originated, but survived in Asia, to which continent it had undoubtedly passed by way of Alaska which at various times has been connected by land with Asia. So it was in Asia that the horse first came to be domesticated.

This horse gave its strength as a supplement to the prowess of man, whose body is comparatively weak, and thereby greatly increased man's power in the world and greatly hastened his development. It was the Aryan races of the old world that first made outstanding use of the horse. It is that race that has developed the ruling peoples of the earth. Had the horse remained in America, had it there been tamed, it might well have happened that the American Indians, instead of remaining the backward race found on this continent when Europeans discovered it, might have developed into a ruling race and might have given civilization an impetus very different from that which has controlled its development. The possibilities that lie in this theory are indicated by the remarkable development among the plains Indians of the West in a single century on horseback. Such Indians as the Sioux, formerly limited by their ability to travel on foot, found their power and enterprise multiplied manyfold when they were given horses, and showed

a remarkable development which found its climax in the
Custer massacre.

It is probable that some wild Tartar back in the middle
of Asia tamed the first horse, mounted himself upon its
back, and found that he had added unto himself its
strength and speed. It would seem that in doing so he
had made one of the discoveries which most greatly affected the developing race. The use of the horse as a
mount, it seems, spread rapidly and soon had largely
covered the eastern hemisphere.

The first outstanding use of the horse was in war.
It increased the range of any people that mounted it
and thus naturally brought them into conflict with other
peoples. When two races met, that which was on
horseback held such an advantage over the other that
it was very likely to conquer it. Thus in the early history
of the world people who rode horses became the dominat-
ing races. The cavalry charges became the wild, irresist-
ible, and terror-inspiring feature in warfare that was
likely to insure victory. Hannibal's horsemen demoral-
ized the Roman legions. The knight, mounted upon
his charger, became the dominant figure in the Crusades.
The followers of Cortez rode horses, which the natives of
Mexico, who fled before them, believed to be strange
creatures, half men and half beast. When East fought
West in the days of Charlemagne, the fate of races
depended largely upon the quality of the contenders'
mounts, and the heavy horses of Charles Martel overcame
the light mounts of the Saracens. Thus do we find the
man on horseback dominant.

The wild horse, which was first tamed in central Asia,

DAWN HORSE OF THE EOCENE PERIOD.

lent itself to becoming a supplement to man largely because it was just the right size to serve that purpose. It was large enough to carry a man, had the speed and the weight-bearing hoof that would enable it to serve this purpose well, and yet it was not too large to be managed and fed by him. It had none of the mental cleverness of the flesh-eating animals, for it did not need to outwit other creatures to get its food, but had merely to crop the grass, usually abundant. It was a simple-minded, pliant creature. For man's primary purpose this Asian wild horse, this ancestor of the shaggy Siberian pony, this creature which remains virtually unchanged in much of the eastern world, was just the right size. It was the sort of horse that man used down through the dawn of civilization, even unto the days of the glory of Greece and of Rome, when, in chariot racing, it experienced a blaze of unsurpassed glory. It was the same kind of horse that Cæsar found among the ancient British when he invaded their islands two thousand years ago.

The first great change that influenced the nature of the horse under its domination by man grew out of that area when knighthood was in flower. The knights of the Middle Ages in western Europe wore heavy armor and burdened their horses with many trappings. Knight, armor, and trappings often weighed four hundred pounds. To bear this weight it was necessary that horses should have greater bulk and strength. So the peoples of western Europe came to breed their horses for ever-increasing bulk. It came to pass that the horses ridden by knights of the Middle Ages, as shown by works of art produced at that time, came to have many of the elements of the draft horse of today. As a matter of fact, they were the ancestors of our modern draft horses, and their sterling qualities of endurance, intelligence, and dependability are to be found in many modern breeds. In the development of these horses, incidentally, breeders of the Middle Ages learned many outstanding lessons in animal husbandry.

Centuries before the Normans thought of breeding a war horse, however, the Arabs had been developing a pure race of desert mounts that in their way have never been surpassed. This horse, like its relative, the barb of northern Africa, living in the pressure of limited grass and water, has been given infinite care, has almost been a member of his master's family. With wide forehead, small nose, arched neck, short back, aristocratic mien, this horse has a look of his own that sets him apart as a creature of a different race from other horses. He is, in fact, of a peculiar race. He has one less joint in his backbone than other horses, and fewer joints in his tail. He undoubtedly is of different origin than other horses— probably comes of an African breed that has been separated for ages from its fellows in Asia.

The glory of the horse as a creature of war in no way diminished until the time of the invention of gunpowder. With the development of firearms and long-range fighting, there gradually came about a condition unfavorable to the man on horseback. So popular a figure was the cavalryman, however, that he clung tenaciously to a place in the armies of the world. With the development of artillery the horse found a new place in battle by furnishing the motive power for larger guns. Then in the Great War the horse found another competition that came near banishing him, in the development, almost overnight, of motor-drawn machinery. The automobile in the Great War quite replaced the man on horseback, the heroic figure which has typified war for three thousand years, and which finds its place in city squares wherever martial valor is commemorated the world around.

But not in war alone has the horse served man well. For thousands of years he supplied man the most rapid and comfortable means of travel by land that there was in existence. To be sure, when the Romans first began building hard roads, the horse failed them, for his feet could not endure these roads. It seems odd, looking back, that five hundred years passed before they hit on the idea of an iron shoe for the horse. This shoe contributed the final element to the success of this master beast of burden.

Nowhere has the horse, as a means of transportation, served a better purpose than in the United States. There was a huge continent, a land of magnificent distances, waiting the explorer and the settler. Soon there was a scattered nation flung along the Atlantic from Maine to Georgia, bound together by the stagecoach. A hundred years ago Massachusetts, Tennessee, Ohio, and other distant commonwealths sent their representatives in Congress to Washington on horseback or by stagecoach.

Then came the migrations to the West—those covered-wagon days when Americans were occupying a virgin empire. Covered wagons were drawn by horses, protected by scouts on other horses. The wild country was prospected by men who rode horses and supplied their

THOROUGHBRED, ARAB, SHETLAND AND SHIRE

needs from the packs borne by other horses and mules. The cattle country could not have been a cattle country but for that romantic figure born in America, the cowboy. So was the horse the most important supplement to the pioneer.

As time passed, different needs have led to the development of different sorts of horses, just as the weight of the armor of the knight led to the breeding of heavy horses in western Europe. There is the great draft horse of our city streets, for instance. One of the most popular breeds of draft horses in the United States is the huge black or gray Percheron, a well-established breed originating in the Perche district of France, a district no larger than an American county. Mounts for knights were here bred six hundred years ago. Crusaders brought back to Perche the best horses they found on their travels. They were selected for peculiar qualities and these qualities have been accentuated as the centuries have passed. So has a choice breed been established, a breed that has been transplanted to America.

The Shetland pony is primarily a draft horse, as is the Percheron. For many centuries, however, it has existed in its inhospitable island home there off the north coast of Scotland, and so great have been the hardships that it has endured and so short has been its food supply that it has steadily grown smaller in stature. It is an outstanding example of the effect upon the size of the horse of the conditions it meets.

Different needs have developed different types of horses under varying circumstances. There is the hackney, for instance, one of the best established, heavy, harness horses in the world, developed in England with a considerable strain of Arabian as its origin. There is the hunter which is a splendid type of animal with speed and sufficient strength to carry its rider over the obstacles encountered in the chase. The thoroughbred is a strain of running horses with a longer record of selective breeding back of it than almost any others, a strain which had its basis primarily in barb and Arabian sires. The standard bred trotting horse of America was based upon this running strain, but has been bred for the development of the harness gait.

Interesting and romantic among the strains of horses that have come into being in the world is the American mustang. Cortez released the first horses in America in 1519, and De Soto in 1539 abandoned a number of horses on the west bank of the Mississippi River. Subsequent Spanish adventurers in one expedition after another released horses. These, with excellent Spanish and North African blood in their veins, became the basis of a wild horse breed which in three hundred years had spread itself over half of a continent. There in the wild state it had gone back to the wild ways of its ancestors. It had, for instance, gone back to olden ways for self-defense. The horse and its cousins, the donkeys and zebras, have a method of fighting peculiar to them, a use of both hind feet in kicking, while the weight of the body is borne on the front feet. In the wild state horses, when attacked by such enemies as wolves, will put their heads together, their heels out, and, by kicking, repell almost any attack. So those horses survived and, when caught and tamed, provided the North American Indians to the Canadian border with an abundance of mounts.

The outstanding trait of these mustangs was hardihood, and when the West was opened up, they became the cow pony, or broncho, which played so important a part in the development of the country.

To the dweller in American cities, even to the dweller upon American farms, it may appear that the horse is being displaced by motor-driven vehicles and that its day is past. The figures on horses raised in the United States, however, do not bear out this conclusion. The number of horses bred in the United States is increasing year by year. Year by year also their quality is improving, because breeders are giving more thought to them, and are realizing more generally that it costs no more to feed a good horse than an inferior horse. Horses are becoming steadily better and handsomer. Agriculture still has the horse as its basis, even in the United States, and is likely to continue to do so. Outside of the United States the reign of the horse is as yet practically undisturbed by the automobile, and its position in Europe, Asia, South America, and elsewhere is as yet no different from that which it held a generation ago.

QUESTIONS

1. Where were the ancestors of the horse when the reptiles ruled the world? Describe the conditions of that time and how the mammals came out of the trees.
2. Trace the evolution of the horse's one-toed foot from its ancestor that had five toes. This toe is called a hoof. What other models of hoofs have you seen? Are they as good as the horse's hoof?
3. To what branch of the animal kingdom does the horse belong? to what order? Does the dog belong to this order?
4. We have seen how primitive man domesticated the dog. Where did the horse come under his control? What effect did the horse have on man's development?

THE HORSE 111

5. Give examples of uses of horses in early wars and of their influence on history. What other examples can you cite from your reading of horses making history? What inventions have lessened the usefulness of horses in war? in peace?

6. Where did the improvement of horses through breeding start? What peculiar anatomical difference is there between Arabian and other horses?

7. How did the horse aid the pioneer in America? Tell the story of the wild horses of the West.

8. Where and how did that magnificent draft horse, the Percheron, develop? What conditions called forth the Shetland pony? the hunter? the standard bred trotting horse?

9. Donkeys and zebras are cousins to the horse and have the single-toed foot. What is the method of defense of this tribe against their enemies?

10. Are the numbers of horses in the world decreasing under the influence of the automobile? Is the quality of the horse improving?

11. How does the horse rank in intelligence with other animals? Which are the more intelligent, carnivorous or herbivorous animals? What is the reason for this? Which must use its head most in making a living, a wolf or a deer?

9

THE BAT

IT is probably true that the strangest and least understood of all the animals in the world is the bat, hanging there, head down, in the tower of the village church.

It is odd, in the first place, that the bat should hang itself up thus by its hind legs when it goes to sleep. One would think that the blood would run to its head. Yet there it is in church belfrys, in the peaks of barns, in deserted houses, in hollow trees, and in darksome caverns all around the world.

When explorers first went into Mammoth Cave in Kentucky they found bats there by the tens of thousands. They fairly covered the roof of the cave, scores of them to the square foot, all hanging by their toes, head down. They had at that time been hanging there for many months without moving, for it was late winter. They were hibernating, sleeping through the cold weather.

Bats are queer from every angle from which you approach them. They are queer, for instance, in this matter of sleeping through the winter. Most animals that hibernate appear quite lifeless, but, upon examination, it is found that they breathe ever so slightly, just enough to show that they are not dead. The bat, however, goes far beyond this; he does not breathe at all. You may fumigate a room full of them for hours with the strongest of poison gases and they will not be injured.

Another odd thing about bats is the fact that, despite their great expanse of wing, they find it difficult to fly, and sometimes cannot fly at all if you put them down on the ground. To get a start at flying they usually have to climb up on something and jump off. They must be launched into the air from some elevation. Left on the ground, they shuffle around to some wall or tree trunk, quite awkwardly, for they can scarcely be said to be able to walk. They reach up with one back foot, get hold, pull themselves up a bit, catch on with the other back foot, and pull up a little more. They repeat this process until they have thus climbed up backwards far enough to launch themselves into the air and fly away.

The bat is odd, unusual, in a class all by itself, in yet another way—because of the kind of wings it has, wings made of skin. It is the only animal in the world that has wings made of skin. Birds have them made of feathers; insects have them made of dry membrane; but the bat alone has them made of live, sensitive skin.

The bat is a vertebrate, an animal with a backbone. Being a land animal, it has developed beyond the fin stage and, like all the rest of them, has four legs. All legs are built on the same plan, so that both its hind legs, which are very short, and its front legs, which are its wings and are very long, have the same upper arm, forearm, and finger formation as have our own limbs.

The wing of the bat, this peculiar foreleg, however, has a most marvelous development, one of the strangest in all nature. Its upper arm is short, its forearm a little longer, and its four fingers are as long as both these taken together. Half the length of a bat, from tip to tip of its wings, is given over to fingers. Its fingers are as long as its body. Its whole wing is but a curious development of the hand. The bat is a wing-handed animal.

Then in the thumb it has worked out a quite different idea. The thumb has nothing to do with supporting the membrane of its wing. The thumb has become a mere claw, a hook at the base of the long fingers. The bat hooks itself on to a tree or a wall with this thumb. The hind legs remain normal legs, somewhat hampered by this wing skin which comes down to the ankles like a petticoat, but with toes ending in sharp, curved claws with which it can hook itself on to any wall or limb and hang itself up for a mere summer day or for a long, cold winter.

The duck and the frog have membrane, skin, between their toes to help them in swimming. In the case of the bat this idea is carried to a most fantastic extreme in the hand. It has developed so much membrane that it can use it in flying. This membrane stretches from the tip of these strangely long fingers, back to its body, runs in a ruffle along that body, out to the very ankles of its hind feet, and on in a ruffle that is supported by its tail. Thus does this expanse of skin run around three sides of its body. It is a thin, rubber-like skin, delicate and light of weight, stretched over a framework of slender and fragile bones, like thin wires, much as an umbrella is stretched over its supporting ribs. It is the only such creation in all nature.

Although the bat flies, it is not a bird. It is, in fact, a mammal, an animal that feeds its young on milk, a

A BAT MUST DRAW ITSELF UP TO SOME DISTANCE ABOVE THE GROUND IN ORDER TO TAKE FLIGHT. HORSESHOE BAT SHOWN ON THE RIGHT

member of that most advanced class of animals which now rules the world. Few people think of bats as being

the sort of animals that belong to the same class as cows
or elephants.   Still insisting on being odd, the bat stands
forth as the only mammal in all the world that actually
flies.   The so-called flying squirrel is merely a glider
which can sail on outstretched membranes from a tree top
to some lower elevation.   It cannot really fly.   Examined
closely the bat seems to be closer kin to mice than to any
other animals.   It used to be called in England a flitter-
mouse.   It has a stocky, fur-covered body, with very
strong chest muscles, since they must furnish the power
for operating those big wings.

But this weird animal has yet another quality, stranger
than any of the rest, that is quite baffling even to the
scientists who have studied it most.   The other mammals,
the most highly developed of all animals, the class of
animals to which man himself belongs, are set down as
having five senses.   They see, hear, smell, feel, and taste.
The bat has all these, but it seems to have yet another, a
sixth sense which none of the other mammals possess.
It can get along very well without the use of the customary five senses—can do such things without them as
can no other animal.

Here is the way in which the peculiar faculties of bats
have been tested.   Modern scientists experimenting with
them, have pasted adhesive tape over their eyes and
turned them loose in different kinds of rooms, just as
you might blindfold one of them and release it in your
own sitting room.   These blindfolded bats would fly
toward the end of the room but turn before they struck
the wall—would not fly into the wall as might have been
expected.   They would proceed across the end of the room,

approach the wall, turn again. They would fly all around the room, but never run into the wall. They would never strike the ceiling, the chandelier, or the lamp.

Experimenters have run wires across such rooms, have dropped strings with weights on them from the ceiling. The bat will wing its way through these mazes of strings and never fly into one of them.

Finally, somebody tried the most puzzling experiment of them all. A bluebottle fly, swift of wing, uncertain in the direction of its flight, was released in this room filled with all these obstructions. The bat is fond of bluebottle flies as food. Still blindfolded, the bat pursued the bluebottle fly, dodging about, changing its course, finally captured it, and devoured it.

Such a blindfolded bat will seek to escape. It will investigate every crack in the room that offers a possibility of getting out. It will hover near the floor where there is an opening under the door. It will approach the keyhole, any knothole, or crack about the window. If those openings are not big enough for it to get through it will not attempt to use them. Open a slit at the window wide enough for it to get through, however, and it will dart out and away, still blindfolded, to its home in the barn.

The bat seems able to get along about as well without eyes as with them. It does equally well when its hearing is destroyed. It cannot be smell, touch, taste that guides it. These objects about the room are not smelled, touched, tasted. Apparently it must possess a sixth sense.

Scientists have long studied this supposed sixth sense of the bat. The generally accepted theory is that bats, living long in restricted, dark places, have found that they needed more than the normal five senses to get about. Man can see pretty well in the twilight, after the chickens can see scarcely at all. A horse, out on the plains, will detect the approach of another traveler before its rider does. A cat or an owl can see better in the dark than can a horse.

This is merely a difference in eyes. There is some light even out of doors on a dark night. The pupils of the eyes of some animals expand more—let in more of that light than do the pupils of other animals. They, therefore, see better in partial darkness.

The bat, however, often lives where there is no light at all. Go a few hundred feet back in Mammoth Cave and there is absolutely no light. If your pupils were as big as windows you would still be unable to see. You could not hear, smell, feel, taste these rock walls. To get about under these circumstances you would need some other sense. The bat, living a part of the time for hundreds of thousands of years in jet blackness, has developed what amounts to a sixth sense.

When a match is lighted in a dark room its light waves travel to the wall and back again to the pupil of the eye. It is their return to the pupil of the eye that brings the image of the wall. When a banjo string is struck it sets up a vibration—sends off sound waves. The ear drum detects the presence of those sound waves. A coarse, loose string vibrates slowly and sends out a low note. A tense, fine string vibrates rapidly and sends out a high note. An instrument may be tuned by tightening or loosening the string.

A milder disturbance of the air would create a mild vibration, a vibration that would not make a noise—that could not be heard. When an oar is pulled through the water, for instance, it creates a ripple. It is a mild sort of ripple, but it spreads and can be detected at the side of the pool. It is turned back at the side of the pool, and a lesser wave returns.

When a bat flaps its wing it creates a similar ripple in the air. That ripple travels across the room, strikes the wall, returns. If the bat had a sense so delicate that it could detect the presence of these air waves, could tell when anything was interfering with them, or turning them back, maybe it could tell where the objects were that were interfering.

This is exactly what the bat does. In the first place its wings, made of sensitive skin as they are, spread out so they can feel the least vibration of the air. Most bats have huge ears in proportion to their size. If a donkey's ears were as big in proportion to its size as are those of some bats, they would be ten feet long. These ears help to pick up these air waves. In front of the ears of many bats a delicate bit of skin sticks up like the tip of a pointed leaf. Its business is to detect vibrations in the air. These vibrations cause it to respond. Certain kinds are known as leaf-nosed bats. This is because they have a similar bit of skin sticking up from their noses. Other bats have other odd-shaped parts to their faces, some of which are grotesque and ugly. Most of them help in this same task of catching air waves.

When the blindfolded bat flies toward a wall, the air waves come back and tell it just how far away that wall is. The chandelier, hanging down in the middle of the room, breaks up the air waves and warns of its presence. A string stretched across the room cuts into the air waves and the bat can tell where it is. There is a point in the wall from which no air waves come back. So the bat knows that the window is open. Finally, the wings of the bluebottle fly make a great disturbance of the air and the unseeing bat is able to follow those disturbed waves and catch the fly.

Bats are to be found almost everywhere in the temperate and torrid zones, but are particularly numerous in warm climates. Despite the fact that they are present in abundance over nearly all the United States they are still little known and little understood by the people. This is largely because they hide themselves away so securely during the daytime and are abroad almost entirely at night when human beings see them, if at all, as vague objects flitting through the dark.

As a mysterious creature of the dark the bat has always been regarded with suspicion, and often thought of as a loathsome creature to be avoided. In fact, it has nearly always been set forth as an evil omen, and sometimes as a fierce demon attacking human beings, except in China where it is highly regarded. A strange belief persists that bats try to get themselves tangled up in the hair of women. This is a thing which may have happened at some time but which is likely never to happen again. There is no more danger of a bat's entangling itself in a woman's hair than there is of a humming bird's doing the same thing.

In certain warm countries, notably from Mexico to

South America, there are bats known as vampires, which suck the blood of other animals. This sort of bat will light upon horses and cows, for instance, cut through the skin on top of the shoulders and suck their blood. Sometimes it makes quite serious wounds. On rare occasions this bat has been known to nip the ears or toes of human beings as they slept and suck their blood. There is no record, however, of the wounds received by men from these bats being serious. Their attacks have happened very infrequently, but have led to the telling

COMMON BAT HANGING AND VAMPIRE BAT IN FLIGHT

of many fantastic stories. The bat has found a place in much fiction in which it has been made to do things it never does in reality.

Most bats in the world live upon insects. The vampire bats are an exception. There are numerous others which are fruit eaters and which occur in both the old and new worlds. The fruit-eating bats of the shores of the Indian Ocean, sometimes called the flying fox, or the fox-faced bat, are the biggest of all the bats of the world. Their bodies may be more than a foot in length and their outspread wings, from tip to tip, may measure several feet. They are so big that they are said to be sometimes hunted for food.

During the day these fruit-eating bats, locally called kalongs, hang themselves, head down, on the branches of trees, often many in a tree. They go forth at dusk to feed upon the mango and other tropical fruits. It is interesting to note that, since they do not live in caves or other dark places, they do not have the sixth sense of the ordinary bats, and have none of the leaflike instruments for detecting air waves. They often do great damage to cultivated orchards. American fruit growers have been very fearful lest this bat should be introduced into the United States, should increase to great numbers, and should injure the fruit industry. Because of this fear very stringent laws have been passed prohibiting their introduction, and so this giant bat is rarely seen in America, even in circuses or zoölogical gardens.

The majority of bats, the common bats of America and Europe, and of all the world for that matter, sleep the day through in their hiding places and come forth in the twilight. Twilight is insect time, and so it is the best hour of the day in which an animal that lives on insects may go foraging for its dinner.

When the bat appears and is seen flitting about streets, lanes, meadows, or water surfaces, it is diligently hunting for insects. Mosquitoes, gnats, flies, and moths are its main source of food supply. In Africa cattle are terribly tortured by flies. Then the bats come out of the woods, follow in the wake of the herds, eat the flies, and bring relief.

The bat's great swiftness on the wing, its ability to turn and twist in pursuit of its prey, make it easy for it to catch most insects. It seizes them with its teeth, which are very sharp and much like those of a cat in arrangement, though much smaller.

This outstanding fact in the life of most bats, of all bats living in the United States, the fact that they live exclusively on a diet of insects, sets them apart as great friends of mankind. The bat is one of the lesser creatures in the world which contributes very ably to the war which man conducts upon that member of the animal kingdom which is his greatest rival and which may some time drive him from the earth—the insect. Man is killing off many of the enemies of the insects, as, for instance, the birds, and is tending to upset the balance of nature, to give the insects a better chance than they should have. There is danger that they may become so strong that they may destroy man's food supply and make it impossible for him to live. For this reason man should come to understand which are his friends in nature— which fight on his side in the struggle to keep the insect enemy in bounds—and should protect them.

The bats, eating only insects, are fighting bravely on man's side. Man, therefore, should kill bats only when special circumstances make it necessary. He should allow them to live in peace in his barn and other places where they do not become a nuisance. Everything that man can do to increase the numbers of bats he should do. Everybody should realize that killing a bat is killing a friend.

This little fluttering shadow that comes past so silently

in the twilight when man may be out with a gun and looking for a shot, may, in fact, be three bats instead of one.
When the mothers come out from the roost they not infrequently bring their babies with them, clinging about
their bodies. These babies hold tight while the parent
flits about, turns, swoops, collects her dinner. They do
not join her in that dinner because they live on milk.
It is very likely, however, that they enjoy the ride. Quick
of wing is this mother, watchful for that other voyager
of the dark, the barn owl, which preys on her kind. The
man with a gun, trying a practice shot, may pick out
such a mother and kill three of his helpful friends instead
of one.

QUESTIONS

1. The bat, in its easy flight, is one of the most familiar creatures of the
   air. Is it a bird? The principal flying animals that are not birds
   are the insects. Is the bat an insect? To what group does it
   belong?
2. The bat is one of the queerest of animals. Name six respects in
   which it differs from all others.
3. How many legs has the bat? The principle on which arms and
   legs are built is the same throughout Nature. The original design
   developed by the amphibians is everywhere followed. How then
   does this wing of a bat, which is its hand, differ from other hands?
4. The birds use feathers to catch the air when they fly. What does
   the bat use? Describe its arrangement.
5. How does the bat feed its young? Can it carry them about?
6. Most animals have five senses. Name them. Describe the strange
   sixth sense of the bat that enables it to go about without sight
   or hearing.
7. Explain the experiment of the blindfolded bat. How can it avoid
   obstacles without seeing them? What special organs has it to
   help it with this peculiar task?
8. This sense of air pressure has been developed because of the peculiar
   life the bat has led. What other instances can you cite of odd
   faculties developed by animals to meet special needs?

9. Do bats entangle themselves in women's hair? Are they otherwise dangerous?
10. Describe the vampire bat and the huge fruit-eating bat of the East.
11. What is the chief food of the bat as it is known in the United States? Why do the bats come out in the twilight?
12. Why should man in his own interest not kill the bats?

Chapter X

THE EARTHWORM

ERE is an odd animal which plays an important part in the welfare of the world by a lowly and unusual means. It renders a great service to mankind, to its fellow members of the animal kingdom, to the plants that grow from the soil, all by the strange device of eating dirt.

This animal with the peculiar appetite, the only animal in the world with this food habit, is none other than the modest earthworm with which one baits one's hook when one goes fishing. So little understood is the earthworm that most people consider it renders its greatest service when it wriggles on a line as a temptation to brook trout. Great would be the surprise of these people should they find out that, aside from this purpose, the earthworm has earned itself a place in the world alongside the cow, the horse, the little brown hen, in the group of most useful animals.

This dirt eater serves its chief purpose by making the soils of the earth more fertile than they otherwise would be. It is Nature's own cultivator of the soil. Through the ages it has stirred up the soil of all the continents of the world more than have the harrows of man. It has worked without ceasing through the centuries, mellowing the earth's surface and making it more productive. It has always added to the world's harvests, and harvests have always been the world's primary interest.

In the average one-acre garden there are fifty thousand earthworms always working. They represent four hundred pounds of brawn busy at the business of soil mixing. They are working in your garden, but they are also working in your wood lot, in your meadow, your cornfield. You may not plan to cultivate the wood lot, but they will none the less keep it stirred up for you against the possibility that you may change your mind. They are mixing soil along the Potomac and out west where the Columbia flows down to the sea. They are at it in teeming India and on the pampas where graze the herds of Argentina. All over the world the earthworms are working in the top soil and their labors make it better for growing the crops of today or tomorrow or that are to feed generations as far beyond us as we are beyond Adam.

The first aid which the earthworm lends to the fitness of the soil comes from the tunnels it runs through the ground. It creates a network of them in the surface soil. Along comes dry weather and the surface soil gets too hard for its operations. It digs deeper, fills the second foot down with holes. Then, as winter comes on, if this is a northern earthworm, it realizes that it must go still farther down, must get beyond the point where there is any chance of freezing. For this and other reasons earthworms may go down four feet, six feet, even eight feet.

These tunnels let the air and the water in. The water softens hard, dry places. The earth deep down becomes

a storage reservoir for moisture that may save plant life when dry spells come. Plant roots follow the worm holes where they might not otherwise be able to penetrate. Thus do they reach new supplies of plant food.

The manner in which this earth-worm, delicate, soft-bodied creature that it is, makes its tunnels through the ground, is very peculiar. If you will examine its soft, jointed body, made up, maybe, of two hundred segments, you will find that it comes to a point at the head. This worm, placed in a pan of soft dirt, will sink its sharp nose into some crack in the dirt. It wedges it between two particles of dirt. It then proceeds to press its inner body into the portion of itself that has got into the crack. This crowds the dirt to the sides. It edges in a bit more, swells itself out, enlarges the tunnel.

Having thus crowded the dirt out and made a passage-way the size of its body, it secretes moisture from its sides. This mixes with the dirt, hardens, and holds the tunnel in shape.

This type of construction, however, will not fit all requirements. Possibly the worm cannot wedge in. There might be a heavy, black, putty-like mass of mud ahead of it. It is necessary to use some other method.

"Very well," says the earthworm. "I will eat myself a tunnel." Whereupon it goes at the wall with its mouth. It fills itself with black mud. Having thus eaten heavily of dirt it decides, by instinct, to make the most of the meal. Since it lives on vegetable and animal matter possibly there is some of this material in this dirt that it has eaten. If so, it will extract it while the dirt cargo is aboard and make use of it.

The earthworm has a crop and a gizzard as has a chicken. The crop lets the swallowed dirt into the gizzard a little at a time. This gizzard is filled with sharp, grinding bits of stone such as the chicken uses, but, of course, much smaller. In this gizzard all the dirt that has been eaten is ground up, worked over. Everything in it, such as decaying leaves, is set aside and used as food. The great mass of it, however, goes on through the earthworm's system and is discarded.

Before discarding this worked-over dirt, however, the earthworm takes it to the surface. Anybody who will take the trouble to look can see piles of it all about, in the garden, along the

THE EARTHWORM LABORING IN ITS TUNNEL

path, at the edge of the pavement. Billions of earthworms every night bring this worked-over dirt to the surface, somewhat as do the ants, and leave it there by their holes, where it can be readily recognized.

The earthworm, however, will eat dirt that has no food for the sole purpose of digging a tunnel.   To prove this, an experimenter filled a jar with fine sand containing no humus, wet it, packed it down. He then put an earthworm on its surface.   The earthworm tried to make a hole by the crowding method, but the sand was packed too tight. It found no crack into which it could get its nose.   It then set about eating its way down. It took it two days on a diet of sand, but in the end it succeeded.

The bringing up of dirt from below is the second great service of the earthworm.   It brings it up from two inches down, from four inches down, from six inches, a foot, two feet, four feet, and piles it at the surface. It does what the plow does, but it goes deeper, works steadier. It is at it everywhere all the time.   One earthworm in one day does not accomplish much, but all of them working through the centuries keep this surface dirt slowly working back and forth.

It would be hard to find a particle of meadow dirt that had not at some time passed through the gizzard of an earthworm.   In doing so it has been ground up; its quality as plant food has been improved.   It is better soil.

A farmer in England once put a layer of quicklime over a meadow which lay there unplowed through the passing years. Six years later he put on a layer of cinders.   Ten years after putting down the quicklime he was digging post holes across the meadow for a new fence. He observed that one inch below the surface all across the meadow he dug through a layer of cinders, and three inches down he passed through a layer of quicklime.

The farmer remembered the quicklime and cinders he had scattered over the meadow years before. It was strange, he thought, that these materials should thus have sunk into the ground. A careful examination showed, however, that they had not sunk into the ground, but that dirt had been brought up from below by earthworms and piled on top of them. Four years later this field was again examined and the quicklime and cinders were found one inch deeper than before.

Careful observation, over long periods, of stones lying on the surface of the ground shows that they gradually disappear. It is an interesting fact that land recently plowed or otherwise brought to the surface is likely to show many small stones. Land long undisturbed offers no pebbles for the boy's sling shot. They have been covered up by the earthworms. Even stone walls, ancient ruins, gradually work deeper into the ground.

In soils where earthworms are plentiful they are believed to bring to the surface an inch of dirt each three or four years. It is fresh dirt that has not seen the sun for a long time, if ever. It is finely ground dirt in form to be used by plants. It is loose dirt likely to drift into the low places. The earthworm has done much through the centuries to smooth out the rough places on the earth's surface. It serves a good purpose everywhere except on the putting greens of golf courses where its little piles of soft dirt are a nuisance. Here it has sometimes been found necessary to pour on liquid poison to get rid of these industrious tunnelers.

The earthworm makes a third contribution to the improvement of the soil of the world. It should be known that this useful animal is a night prowler. Like the stealthy rodents and the flitting bat it is abroad chiefly when conventional people are comfortably tucked in their beds. By day the earthworm is likely to sleep just underground with its nose very near the opening to its tunnel. If you suddenly scrape away the earth it has brought up you are very likely to see it making a quick retreat.

When the light of day is gone, however, it comes forth. It does this for several reasons. Then it may meet other earthworms. Nighttime is its frolic time, for there in its tunnel by day it is all alone. Then, also, it may get a bit of green food to vary its monotonous diet of dirt. Further, it is safe at night, for the birds that so delight in eating earthworms are on their roosts.

Yet the earthworm is a modest sort of forager when it goes leaf hunting. It must be because it is not very well equipped for careering in the great upper world. In the first place it cannot see at all. It has not so much as the semblance of an eye. It does not need eyes down below and they would not be much use to it when it comes to the surface, since it nearly always does this by night.

Neither can the earthworm hear, since it has no ears. Its body is sensitive to light. It can tell the difference between light and darkness. Its body is also sensitive to vibration. It can tell when a workman is busy near by with his spade, or when the mole, its arch enemy, is tunneling in a near-by bit of earth. These senses are entirely aside, however, from those of sight or hearing.

The earthworm has no sense of smell. It has no lungs, but absorbs air through its skin by a process that may be called breathing, but which really is not.

There are left to it but two of the five senses, with which it must do the best it can. These are the senses of feeling and of taste. With nothing else to guide it, this modest, lowly animal must earn its living, must render its hugely important service in the world.

When the earthworm comes to the mouth of its tunnel in the nighttime it is very likely to be looking for garden truck for its supper. In its hunt for it, as a poor blind creature, it must be careful not to lose its way. It takes a great deal of pains to keep in touch with its home in the ground. Men, when they go into unknown caves in the underground world where the worm is at home, often tie string to some object at the mouth of the cave and unwind it as they go, that they may follow it back to the surface. Acting upon the same principle, the earthworm, when it comes out into the air, ties itself back to that underground world which is its natural abiding place. It does not carry a ball of string, but its own body may be so lengthened as to serve practically the same purpose. This blind and deaf creature with no sense of smell leaves its own tail snugly anchored in the mouth of its tunnel. Thus made secure it reaches out with its body in all directions in search of leaves.

With its toothless mouth it may seize small leaves or blades of grass and drag them into its lair. When a leaf

is too big to be handled in this way or when the worm
cannot get hold of it with its mouth it resorts to a different method of attack.   It draws the pointed part of
its body back inside its bigger rings as the tip of the
finger of a glove may sometimes be pulled back inside
the finger.   Thus it changes the forward end of its body
from a pointed object to a very blunt one.   This blunt-
ended body, in fact, becomes a suction cup.   It presses
it against the leaf and withdraws the head farther within
the body which causes the suction pump to take hold.
With a leaf thus grasped it pulls its body back into the
hole and the leaf with it.

Vegetable matter that is thus taken underground by
the earthworm serves two purposes.   In the first place,
the worm may cover a leaf with acids secreted from its
own body, thus softening it and making it possible to
eat it even with a toothless mouth.   The great majority
of this vegetation, however, is used for the purpose of
blocking up the earthworm tunnels and thus making
them secure from intrusion.

Whatever the object of the earthworm in bringing
this vegetable matter underground, it serves a valuable
purpose by being converted into fertilizer.   Farmers plow
under green crops, scatter straw upon the land, and take
great pains in one way or another to introduce vegetable
humus into the soil, because they know that this vegetable matter will cause that soil to yield better crops.
This modest little assistant gardener for all the world
has been working steadily at getting humus underground
for millions of years.

The benefit that the earthworm has done by running

tunnels through the surface soil, by constantly bringing up subsoil, by constantly taking vegetable matter below the surface, is one that is beyond the power of man to estimate. We might conceive that this constant service of the earthworm through the centuries has made the earth twice as productive as it otherwise would be, that it has made the soil fifty per cent, ten per cent, more productive. We might even estimate that the earthworm has made the soil produce a modest one per cent more than it would have produced otherwise. There is no way to determine the fact, but even this one per cent increase in the productiveness of all the soils of all the world would have been a stupendous service to man, to animal life, to plant life, would have been a service such as few animals have ever rendered. A one per cent addition to the crop of the United States alone would amount to a hundred million dollars a year.

It may seem odd that this earthworm should be referred to as an animal. We must, however, go back to the fact that all living things which are not plants are members of the animal kingdom and are, therefore, animals.

The first great division of the animal kingdom is into two parts: the vertebrates, which have backbones, and the invertebrates, which have no backbones.

The more highly developed animals, the familiar animals like cows, cats, birds, fishes, belong to the backboned groups. It would be a great mistake, however, to overlook the great groups of animals that are without backbones, groups of much more variety and stronger in numbers than are those with backbones.

The invertebrates are divided into a number of phyla, or branches, some of which, as the starfish, are not very important, and some of which, as the jointed animals, including the insects, are very important. The principal phvla are as follows:

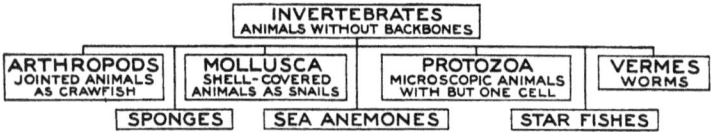

INVERTEBRATES ANIMALS WITHOUT BACKBONES			
ARTHROPODS JOINTED ANIMALS AS CRAWFISH	MOLLUSCA SHELL-COVERED ANIMALS AS SNAILS	PROTOZOA MICROSCOPIC ANIMALS WITH BUT ONE CELL	VERMES WORMS
SPONGES	SEA ANEMONES	STAR FISHES	

The earthworm, of course, belongs to the Vermes, or worm, branch of the invertebrates. This is a small branch with not nearly as many members as one who had not studied it would think, for, strictly speaking, there are not many worms. To it belong the earthworm, the sandworm, which is found along beaches, the tapeworm, which lives in the intestines of man and animals, the leech, which lives in the bottoms of brooks, and a few others. They are the only true worms. They are the members of the animal kingdom which have soft, jointed, uncovered bodies, and which retain that sort of bodies all through their lives.

It is quite the custom to speak of the silkworm, of the measuring worm, of wormy apples, of worms that may get in the cornmeal, of bookworms, of worms that may be found in great numbers in manure piles or other filthy places. None of these, as a matter of fact, is a worm. The silkworm is but a moth in the undeveloped stage, as is the larva which bores into the apple core. They are but undeveloped insects, baby moths, and belong to the order of insects. The disturber of the cornmeal, the borer through books, are but young beetles which will soon have six legs and shells on their backs. They are not worms. The wrigglers in the manure piles will soon be buzzing around the kitchens as houseflies, those villains of the insect world which, through spreading disease, have killed more people than all the wars.

The earthworm is, however, one of the true worms, the good citizen of its kind, an important contributor to the welfare of the world.

I have here spoken of "the earthworm" as an individual, have described its habits as those of an individual. This should not be taken to mean that earthworms are all alike. They are no more so than are all men. There are probably as many varieties of earthworms as there are of men. The fisherman knows that there is a brown and yellow striped earthworm to be found about such places as manure piles, a sort of slum dweller of its race, that has a quite different appearance from the pinkish fellow found in the rich soil at the edge of the woods.

By the same token there are in this country native American and immigrant earthworms. The peoples who came here from Europe brought plants with them and earthworms in the roots of those plants. The worms often prospered more than did the plants. The earthworm which piled the dirt on top of the lime and the cinders in England and which, by the way, is one of the best soil mixers of them all, did not exist in America until it was brought here by Europeans. As long ago as the last decade of the last century it was hardly to be found, for example, round about Urbana, Illinois. Yet a generation later it was abundant and working industriously. It has shown itself one of America's desirable immigrants.

QUESTIONS

1. Have you ever examined an earthworm to see how it is put together? Have you ever seen one in its home? As a part of this
   lesson, find the right sort of soil, dig up some earthworms, and
   examine them.
2. What is the master service that the earthworm renders the world?
   Name three ways in which it improves the soil.
3. Describe two ways in which the earthworm makes its tunnels.
   Why does it eat dirt?
4. Point out some of the ways in which a measure has been secured
   of the amount of dirt the earthworms bring to the surface. How
   does the earthworm help to keep the ground smooth?
5. Owls, bats, and toads are night prowlers. Why does the earthworm join this group that shuns the light?
6. Which of the five senses that are a part of the equipment of most
   animals does the earthworm have?
7. How does the earthworm keep from getting lost? What food supplies does it gather above ground? How does this fertilize the
   soil?
8. The corn crop in the United States alone amounts to two billions
   of dollars a year. If the earthworm has added ten per cent to that
   crop alone, how much profit has it made for the farmer?
9. Is the earthworm an animal? Is it a vertebrate or an invertebrate?
   There are seven branches of the invertebrates. To which of these
   does it belong?
10. What is a worm in the proper sense of the word? Why is a silk-
    worm not a worm? Are caterpillars worms?
11. We have seen the work that earthworms do in England. Are Amer-
    ican earthworms just like those in Europe?

CHAPTER XI

THE COW

TO look at a drab bale of hay, lying there so uninvitingly on the floor of the barn, one would little suspect that there was wrapped up in it a bucketful of foaming, creamy milk, the most perfect food for human consumption in all the world.

Yet stoke that hay into one of those highly developed machines which man has caused to exist by virtue of taking thought—into the old spotted cow yonder in her stall—give it a bit of time, and, presto! She gives you the best food known to man.

The cow is a milk factory. You furnish her with the raw materials and she turns out the finished product. Not only does she do that, but she is in touch with the spirit of the times and has gone in for quantity production. Give her the makings and she will fill forty quart bottles for you in a single day. In two weeks she will yield her weight in milk. In the course of a year she will produce the weight of a herd of twenty-five cows of her kind. She is the most remarkable of living, breathing factories. There is nothing else like her. She has a virtual monopoly in her field. She manufactures an article for which the demand in this world is twice the supply. There is not enough room for enough cows to produce enough milk.

We are likely to accept the cow as a commonplace without ever stopping actually to consider her, without ever setting her out all by herself and taking a good look. Familiar as she is to us, we are not likely ever to have even wondered how she came about. We have

IN ONE MONTH A COW CAN PRODUCE TWICE HER WEIGHT IN MILK

not realized that she is a marvel, one of the wonders of the modern world.

The cow converts hay and grains into milk at man's order. At will, however, he may switch the order and get an entirely different product—beefsteak. The cow, under a different treatment, is a meat factory. Give her this hay and grain under a different set of circumstances, and she will become a mountain of meat, a ton of it there on four legs ready for the butcher.

The cow did not come to this fine adjustment all by herself. In the beginning she was a bony sort of creature running wild, with an occasional cupful of milk for the nourishment of her calf. Think how quickly the wolves would have pulled down the stall-fed prize winner at the State fair and how sumptuously they would have feasted! Think how helpless the dairy cow with her huge udder full of milk would have been in her old home in the primeval plains of the old world!

The cow was a wild thing with certain traits that might be developed to man's advantage just as the tiny "love apple," once thought to be poison, had within it the possibility of being developed into the tomato. Man, beginning to think, sought a way to increase the usefulness of the cow. He ended by accomplishments that are nothing less than marvelous.

The ancestors of the cow in those days when the world was young are not so distinctly known as are those of the horse. It is evident, however, that they lived in the tree tops with the ancestors of the horse and had toes with which they could cling to the branches, at the time when huge reptiles ruled on the ground.

What is now America's great plains area was then a lowland, a morass, teeming with vegetable and animal life. This can be definitely established by an examination of the rocks of that time. This lowland area was gradually elevated through the action of the earth's surface, was changed from a lowland to a highland. Thus it was made an unfit place as a home for the huge reptiles of an earlier time, and they ceased to live there. The mammals came down out of the trees and gradually changed into forest animals that walked upon the ground. Then, as the forest changed into grassy plains, they adjusted themselves to these new conditions, entirely lost their abilities as tree climbers, and steadily developed those qualities which fitted them to survive out there in the open. At other points on the earth the same sort of changes came about.

Slight differences in these creatures that came out of the trees caused them to develop along different lines. The horse family, with its single hoof and its better shaped body, built up one group, while the cow family, with its split hoof, came to make up another group. In this family there were different branches which grew steadily farther apart until we have today the cow, the deer, the elk, the moose and, a little further removed on one side, the camel, and on the other, the hog.

Rock remains show that the horse made its change on our own American plains from a forest dweller to a creature very much like what it is today. It disappeared from those plains, however, before the coming of man. It had crossed over to Asia and was there first tamed by man. We do not know so well just where the cow developed. The sort of cows that came to be used by man seem to be natives of Europe and Asia.

It was unfortunate for the man life of the American continent that the animals that might have helped it develop more rapidly died out before man came. The fact that the race did develop most rapidly in Asia may have been due, to a large extent, to the presence there of horses, sheep, and cattle that lent themselves in usefulness to man. The only creature that survived in America which

PRIMITIVE PROGENITOR OF OUR MODERN CATTLE

is closely related to the cow, and which might have been
tamed, is the American bison or buffalo. The buffalo
has been occasionally tamed since white man came to
American. It will pull a plow as does the ox, but if it
gets thirsty and wants to go for a drink, it goes. It
cannot get the idea of being restrained by man so well
as does the ordinary ox.

The animals from which the modern domesticated
cow of western nations came were wild cattle living in
western Europe. These wild cattle were tamed early
in the history of the human race. Among the peoples of
western Europe that may be traced farthest back were
the lake dwellers of Switzerland, living in huts built over
the water for protection from wild beasts. These lake
dwellers were among the first people of Europe to begin
to emerge from savagery. Excavations from the ruins
of their dwelling places show that they grew crops and
maintained their flocks two thousand years before Christ.
Among the relics of that early day are to be found cattle
bones which indicate that these animals played an important part in their lives.

In Biblical days, before western European people had
any civilization to speak of, the wealth of kings such as
Solomon lay in their flocks. Flocks* and herds had
developed much earlier in Asia and Africa. When the
Greeks first began to use metal and stamp it into coin,
they placed a bull upon this coin as a symbol of value.
The keeping of flocks of sheep and of cattle was virtually
man's first industry. This industry preceded the time
when he began to grow crops and was a part of his life
when he was a wandering nomad without any settled home.

It is interesting to conceive of the possible occasion when some old savage got the idea that it would be better to raise his own herds of cattle than to depend upon the chance of the chase for his meat. It is interesting to imagine a time, probably thousands of years later, when some mother of the tribe died, leaving a young baby, and its father thought of getting milk from the cow to keep it alive. Then there was the first forked stick drawn by a bull to scratch the soil. Finally, much later, a great man of his day first thought of the advisability of killing for meat the poorer members of the herd and saving the better members to carry on their race. This was the principle of selective breeding.

It remained for the people of western Europe, probably because of peculiarities of character, to solve the riddles of stock breeding. The genius of that people which conceived the possibility of developing a bigger horse to carry its armored knights into battle also saw, a few centuries later, the vision of developing a race of cows that would supply it more abundantly with milk. That people began the deliberate application of the principle of selective breeding to cattle.

The most striking practical results of the application of that principle are to be found in the past two or three hundred years. There is the case, for example, of the Holstein cow, the greatest milk producer of the world. The Holstein cow is the product of taking thought, of selecting from the herds those cattle that give most milk and making them the parents of the next generation, and repeating this selection generation after generation through the centuries. In this way each generation of cows gave a little more milk than that which had gone before. It is because this principle of selective breeding has been applied there in Holland that the Holstein cow is today the marvel of her kind and produces her weight in milk in a brief two weeks.

There in the Channel between England and France are to be found the islands of Jersey and Guernsey. Though they belong to Great Britain, these islands used to be more French than English and so they got their idea of cattle breeding and their first cattle from Brittany and Normandy in France. By selection they established types of cattle and, because they were islands shut off from the rest of the world, it has been easier for them to keep their strains pure.

In Jersey, for instance, it has been against the law since 1779 to import any animal except for slaughter. The Jersey cow as a producer of milk is the result. As the Hollanders selected their cattle for producing quantities of milk, so the people of Jersey have selected theirs with the idea of quality—richness. The Holstein cow is known all around the world as the breed that produces most milk and the Jersey cow is known as the breed that produces milk of richest quality. These are the two aristocrats of the dairy world.

It is an odd thing, showing the character of peoples, that the Britishers should have taken the lead in the development not of milk cattle, but of beef cattle. The Britisher has always been a beef eater, and in the development of his herds he has built up the three dominant beef-producing strains of the world.

There on the River Tees in northern England about

the time of the American Revolution there began to take form a breed of cattle known as the Teeswater cattle, later called Durham cattle, and still later, Shorthorns. The Shorthorn is the prime beef animal of all the world.

Down in middle England at about the same time there was a breed of red and white spotted cattle with long, drooping horns. The model of this breed came to be an animal with a white face and a red body very definitely marked and, through selecting only white-faced, red-bodied cattle to become parents, this marking came to be established. The breed was given the name of Hereford since it came from Herefordshire. It likewise was a stocky, beef-producing animal, but active and resourceful from foraging in the hill country.

In Scotland there was developed the Galloway, a shaggy, black animal, likewise yielding great quantities of beef. It has no horns because the Scotch selected the individuals that tended to be hornless and bred them together.

Wherever in the world progressive stockmen have wanted to improve the breeds of cattle raised for beef, they have sent back to England for Shorthorns and for Herefords and to Scotland for Galloways. Creatures of these bloods may be seen today ranging the pampas of the Argentine and the plains of Australia. They have remade the breeds of the great open spaces of the United States which, a generation ago, were one of the fittest places in all the world for the development of vast herds.

Spanish colonists, nearly four hundred years earlier, had released their cattle to the south, and these had overrun the plains. To be sure they had lost quality, had become scrawny, wiry, long-horned creatures whose frames carried little meat. But they were hustlers—they could get along under adverse circumstances. With these wild cattle the master beef breeds from Britain have been crossed, and a new race of range cattle has resulted.

Odd differences appear when one stops to consider these tribes of dairy or beef cattle. The producer of milk, for instance, has certain peculiarities of bodily form. The organs which for her are all important are two, her stomach and her udder. Her stomach is vital, for it must digest great quantities of food and then must convert it into milk. Her udder is important, for it is there that the milk must be held. Four or five gallons of it twice a day must be stored there until the master comes along and removes it. It makes little difference about the rest of her form as long as it is a frame that takes care of these two organs. So the milk cow is likely to be sharp fore and aft, narrow at the back. It is particularly noticeable that she is wedge-shaped in front. Her body comes almost to a point at her thin neck.

With the beef animal the problem is entirely different. Here the breeders have paid most attention to the development of a framework, not too heavy in bone, on which may be hung the largest quantity of meat. The body of a beef animal is a block shaped like a brick set on edge. The back is broad and flat; the shoulders and the hips wide and square. Through the centuries beef growers have selected animals of this sort, that were blocky and heavy, as the parents of their herds. Those that were not of this form were eaten young and had no calves. Thus have been developed beef prize winners that have weighed up to and above two tons, four thousand pounds. This is again a result of selective breeding.

The hog is a split-hoofed animal like the cow. It is related to the cow to a degree that might be likened to a distant cousinship. The hog is closer kin to the hippopotamus than it is to the cow. The hog has the same sort of stomach that the hippopotamus possesses, while members of the cow family have a very odd and complicated stomach, different from that of any other animal. They have, in fact, four stomachs in a row. Animals with these compound stomachs are called ruminants.

The ruminants are animals which, in the beginning, had to get their dinners in a hurry, for, when they came out in the open to graze, they were in danger of being caught by the flesh-eating animals. Cows, sheep, deer, antelopes, camels, and llamas are ruminants. They live by cropping grass, and they are built to do it quickly. They run their tongues out to gather in a clump of grass, something as an elephant uses its trunk. They close the front of the mouth on this grass, holding it as though in a vise. With it thus held they thrust their heads forward. They eat "away from themselves." They have no upper teeth in front, but a row of sharp lower teeth. This forward thrust half cuts off, half tears off the clump of grass.

All of this is quite different from the horse's method of eating. The horse cannot gather in the grass with its tongue, but it has an upper lip that can be used for lightly grasping it. The horse, also, has teeth both above and below. It gets the grass between these two rows of teeth and crops it by pulling its head inward. It eats "toward itself."

The ruminant, in a hurry to get back to cover, bolts this grass without chewing it. The grain goes into its first stomach. When the animal has reached safety it lies down placidly and its stomach begins to perform its peculiar service. The food goes from the first into the second stomach which is made up of small compartments that divide it into "cuds" that are a neat size for chewing. These it sends back to the ruminant's mouth at the proper time to be ground up. The animal chews them at leisure and reswallows them, and they find their way into a third, and finally into a fourth stomach, and are digested. Only ruminants eat in this way. Because of their thoughtful attitude while chewing their cuds it has come about that one who thinks quietly is said to ruminate.

So simple is the manner in which the grass-eating animals get their living that it has never been necessary for them to develop the intelligence essential in the case of the flesh eaters which must outwit and catch their prey. The chief concern of the ruminants has been to avoid their enemies, and to do this they have held to the policy of showing a clean pair of heels. They have run away. To them, however, watchfulness has always been necessary. The carnivorous animal goes about looking for game that he may capture. He is a hunter. His eyes, to serve his purpose, are in the front of his head. The rôle of the herbivorous grass-eating animal, on the contrary, has been that of watching out for the flesh eater that might be trying to slip up on him. He has, therefore, found it necessary that he should be able to see all around—hear all around. His eyes and ears, consequently, have been placed at the sides of his head that they may the better stand guard. Thus does Nature adapt the organs of her creatures to their special needs.

Man has taken some pains to develop the intelligence of the horse. It was necessary to make him more useful. In the case of the cow, however, placid stupidity has been an advantage. There was no need for the cow to be clever. Under the influence of man she has grown more stupid than she was in the beginning. Her rôle was merely to eat grass and convert it into milk or meat. For this she needed little intelligence. It has been the same as in the case of the silkworm moth which has been a domesticated animal for many thousands of years. The moth that was the least active was most easily handled. So, as the centuries have passed, man has bred a silkworm moth which, though a creature of wings, scarcely moves from the place he puts her there on a piece of clean muslin, until she has served her purpose by laying a cluster of silkworm eggs.

The cow, despite her placidity, however, has retained some of the instincts of her one-time wild life. When her calf is born, for instance, she will hide it quite cleverly in the grass. The calf, by instinct, will lie there, scarcely moving for twenty-four hours or more, until the mother returns for it. Yet the mother instinct of the dairy cow has largely disappeared under the influence of the dairyman. For so many generations has her calf been taken away from her that she no longer grieves for it. She accepts without complaint the fact that it is to become an institutional child without the influence of mother love. She has given up the love of her young and in return has received comfortable stables and unending and unstinted supplies of food. As a gourmand she has developed for herself a life of unending satisfaction.

Among the bulls the instinct for combat still lives— a survival of the time when the most powerful male of the herd appropriated the cows unto himself, stood guard over them and gave battle to any other like bull that might appear. This manner of life causes cattle to be ranked as gregarious animals, animals that tend to sociability, that flock together, as opposed to animals, like the bears, that keep to themselves.

The bull is still monarch of the herd. Today if two old dairy bulls, gentle as dogs, each with centuries of breeding back of him, be thrown into the same pasture, there is bound to ensue a combat for the mastery, a fight which for force and ferocity, as a contest of titanic strength, is almost without an equal on this earth.

And cattle on the march! There again is an example of the inheritance from the days of the open plains, when the herd went in quest of new grazing grounds. Though it sets out to cross the small meadow back of the barn, it falls into lines as solemn and orderly as any formal procession. This is a staid pilgrimage, and each member must play her rôle with dignity.

It is on such pilgrimages that cows give evidence of a quite marvelous instinct, despite their stupidity. They lay out their paths with the accuracy of an engineer. They choose the shortest distance on the best grades between two points. Roads across great plains have followed these cow trails, and engineers of later generations have not seen fit to change them. Wall Street was first a cowpath laid down by black and white Dutch dairy cows brought over by the early settlers. Builders of railroads followed buffalo trails through the easiest passes of Pennsylvania mountains. When an army post was established where Los Angeles now stands, the covered wagons followed a cowpath which became a wagon road and is now Main Street, in that city.

In a material way the cow has served man better than any other creature. She has drawn his loads and his plow. She has yielded him milk, cream, cheese, and butter. She has produced for him more and better meat than all the other animals combined. She has grown steadily in her usefulness and is still growing. Every year he finds new possibilities of using her to his profit. Only a generation ago, for instance, the cow was profitable only in the grass-growing season and was an expensive burden to be carried through the winter. Now the dairyman, under improved feeding methods, makes profits from her, there in her stall, every day in the year. The noontide of her glory is not yet reached; while it is suspected that the day of the horse and the dog is on the wane. She is unequaled as a living organism that produces always wealth today where yesterday there was none. She is the provisioner of the cupboards of the world—man's most productive dumb servitor.

QUESTIONS

1. We have learned from the rocks that the horse family developed in America, but later died out here while surviving in Asia. What can you say of the development of the cow family? What is America's nearest relative to the cow? Does it lend itself to domestication?

2. Where were the ancestors of our cattle first tamed? What was man's first industry? Tell of the conditions under which herding probably developed.

3. Where did selective breeding begin? What are some of the well-known strains in cattle that have resulted?

4. How long does it take a Holstein cow to produce her weight in milk? How was she developed into performing this miracle?

5. What can you say of the livestock contributions of the Channel Islands? The chief purpose of a herd may be the production of milk or it may be the production of beef. Tell of the development of the beef breeds.

6. Observe and make a list of whatever cattle you may see for a week, note their peculiarities, try to determine the breeds to which they belong. Set down the countries from which their ancestors must have come.

7. What are the peculiarities of form of the milk cow? of the beef animal?

8. What is a ruminant? Describe the peculiar manner of eating of the ruminants. How did they get the habit of eating in this odd way?

9. How do these grass-eating animals rank as to intelligence? Is the cow intelligent? What is the difference in the arrangement of the eyes and ears of grass-eating and of flesh-eating animals?

10. What instincts handed down from a time when she was a wild animal may be observed in the cow? How may the cow become a guide to the engineer? Where has she left her mark on the map of the nation?

11. The cow contributes more to the well-being of man than any other animal. Name all of the products you can that he gets from her.

CHAPTER XII

THE WHALE

THE largest animal that has ever lived in this world still lives. It is none other than the whale, that huge beast which wallows in the deep seas of all the world, occasionally showing itself here and there to mariners who thread any of the mighty oceans.

It comes about quite naturally that the biggest of animals should be one which lives in the water. Out there in the ocean there is plenty of room for the big creature to roll about without colliding with other objects. There in the water, also, it is suspended by pressure all about it and its great weight needs nothing upon which to rest. There is no problem of operating legs that are sturdy enough to hold it up. It need not consider such questions as the firmness of the ground upon which it walks and whether that ground will bear its weight or whether its feet will sink into the mud. There in the ocean also are inexhaustible supplies of food, and so big an animal must have great quantities of breakfast, lunch, and dinner.

There was a time on land when the vegetation grew much more rankly than at present, a time from which has come down only the sequoia, or big tree of California. In that age of rank vegetation huge reptiles thrived, animals many times bigger than any that walk the earth today, reptiles, the remains of which are to be found in the rocks of long ago. The land, however, never produced an animal that was as big as the whales which now swim in the waters of the ocean.

The question of just how big is the biggest whale is one which remained long unanswered. So impressive is the bulk of these monsters that stories of their size are likely to be fantastic and exaggerated. Since there are no scales of the deep on to which it may be led, the weight of a whale is nearly always a matter to be guessed at, and it is usually set down as being greater than it is. It was in fact not until 1903 that anyone ever actually put a whale on the scales and weighed it. In that year a blue whale, the kind of whale that grows larger than any other, was actually cut to pieces on the coast of Newfoundland and weighed, piece by piece. This whale was seventy-eight feet long and thirty-five feet around the shoulders. It was a fairly large whale, but by no means of a record size. On the scales it was found actually to weigh sixty-three tons, to be as big as one hundred draft horses, as big as a dozen elephants. A blue whale elsewhere in the world has been measured and has been found to have a length as great as eighty-seven feet. Such a whale probably weighed seventy-five tons. There may be blue whales in the world that are one hundred feet long and that weigh one hundred tons. Such a whale, the biggest animal that ever lived, would weigh as much as all the inhabitants of a town of sixteen hundred people.

Of the whales of the world there are two major divisions. There are the whalebone whales, which have no
teeth and feed on tiny animals that live in dense swarms
in the ocean; and the toothed whales which prey chiefly
upon fish and squids. Most of the big whales belong tó
the whalebone group and use this whalebone, for which
they are hunted, in a very odd way in getting their food.

Whalebone, or baleen, grows in the roof of the mouths
of these whales. It consists of long strips shaped very
much as are the stems of the leaves of palm trees. The
biggest piece of this baleen may be fourteen feet long,
while an average length is four or five feet. Big pieces
of whalebone may be eighteen inches across at the base,
coming to a point at the tip. These sheets of whalebone
have a broad, fibrous fringe along their overlapping edges.
There is a row of this fringed whalebone on each side of
the mouth which forms a very perfect strainer.

The whale, as it swims through the ocean, where its
minute food lives so abundantly that the water often has
a brownish tinge, opens wide its mouth, and allows great
quantities of water to pass into and through it. As the
water flows out the sides of the mouth the small animal
life, or brit, it contains is retained and swallowed in enormous quantities, literally by the barrelful. It repeats
this process until it may have taken aboard something
like a truckload of food and calls it a meal.

This great blue whale, often also called the sulphur-
bottom whale, is of the baleen type. When whale hunters succeed in killing one of them they are likely to cut
about eight hundred pounds of whalebone from its mouth.
Getting the whalebone is one of the elements of profit in

the capture of the whale. Whalebone is cut into many
forms and used in manufacture. It is familiarly seen in
boning that bends easily yet holds clothing in shape, or
as the cores of flexible whips.

The second element of profit from the whale lies in
the whale oil that is secured from its carcass. This comes
chiefly from the layer of fat averaging a foot in thick‑
ness with which whales cover themselves just beneath
the skin as a means for keeping warm. This layer of
fat is a blanket. It is a nonconductor of heat; it will
keep the whale warm in the coldest water. Whalers strip
this fat from the bodies of their victims in an interesting
way. They peel the whale much as an orange might be
peeled. They start a strip of fat at the head, running
around the beast, then they attach a donkey engine
aboard ship to the end of it. As the engine pulls they cut
the fat along the attached side with sharp spades. As the
strip is pulled off, the carcass of the whale turns over and
over as would a skein from which yarn was being un‑
wound. Finally the fat is stripped off clear to the tail.
It is promptly fried out. The oil yielded by this process
from a good-sized whale may amount to one hundred
barrels.

The bowhead or great polar whale is not so big as the
sulphur-bottom, but is more profitable to the whale
hunter. It has an enormous head which grows much
greater quantities of whalebone than does that of the
blue whale. Eighteen hundred pounds of whalebone is
no uncommon harvest from a bowhead, and as much as
three thousand pounds have been secured from a single
individual. The bowhead is a ponderous, bulky, ugly

whale, sixty feet long, whose huge head occupies more than a third of its body. The skull of that head is six feet thick and weighs five tons.

Some of these whales have hearts which weigh a ton and tongues which weigh three or four times as much. The finback is a huge fellow, next to the great blue whale in size. The humpback is a whalebone whale well known to hunters. The California gray whale, about fifty feet long, is frequently seen, as is the sei whale of Japanese waters. These are the chief remaining whalebone species.

FROLIC OF A SCHOOL OF SPERM WHALES

A quite different animal from the whalebone whale is the sperm whale, the only big whale with teeth, a beast of prey which lives on huge sea squids and fish. Instead of whalebone in its mouth it has a row of large ivory teeth looking like huge pegs set along its jaw. The loss of value of this monster to the whaler, as compared to the baleen producer, was formerly made up by a prize it carries in its head. It has a huge head shaped like a big

12

box, and inside it is a cavity filled with spermaceti, an oil of superior quality. When refined it yields a product highly prized for oiling fine machinery. There used to be no oil so good. Modern science, however, has developed a mineral oil from petroleum that has pretty well killed the demand for spermaceti. This tank in the whale's head was formerly, however, a storehouse of wealth. When the sperm whale was captured it was only necessary to open up its head and dip out this oil with buckets, a single individual often yielding a dozen barrels of oil.

Scientists have debated much upon the purpose of this tankful of rare oil to the whale. The general conclusion is that it carries this tank as a reserve food supply for use in times of scarcity. Possibly there is a season at which it does not feed, but lives upon its reserve as the bull seal lives upon its reserve fat during the breeding season, or as the camel lives upon the fat stored in its hump when famine time comes.

Another very peculiar product of the sperm whale is ambergris, one of the strangest articles of commerce in all the world. The sperm whale is the sole source of the supply of ambergris for the world. Ambergris is itself the raw material for a kind of oil. It is used chiefly in the manufacture of fine perfume. It has some peculiar quality that causes it to retain the odors of the perfume as will nothing else that man has been able to find or make.

Yet this ambergris is not a natural product of the sperm whale. Not one of them in a hundred yields any ambergris. Only sick whales, in fact, yield it. In deposits of ambergris are nearly always found the beaks of cuttlefish, and it is thought that the presence of these in the intestines of the whale causes the ambergris to form. It may be deposited around these cuttlefish beaks to keep their sharp edges from doing any harm. The ambergris is sometimes found floating at sea or washed up on beaches. An ounce of it is worth $15 and as much as $60,000 worth of it has been secured from the intestines of a single whale.

Whale hunting is a very ancient industry, first followed in the Bay of Biscay a thousand years ago. It was whale hunting that brought the first Norsemen by way of Greenland to the American continent. The pursuit of whales has been a calling filled with romance through all the development of man and his going down to the sea in ships, finding its climax among the fishermen of our own New England coast and its flower during the middle of the last century. In its modern development, however, whaling has lost much of its romance— has become a venture in which the whaler takes little chance and in which the whale is merely slaughtered.

In the old days the whalers went forth in small boats, threw harpoons with ropes tied 'to them into the whale, and were taken on wild rides full of danger of death over the ocean. They tired out their victim, maybe in a few hours, maybe in a day, then finally lanced and killed it. In modern whaling a ship comes alongside its unsuspecting victim wallowing on the surface, and fires a huge harpoon into its body from a gun, a harpoon which carries a bomb that explodes in its vitals and rends and tears it. Even though it is not killed immediately it has the weight of a heavy vessel to drag, a vessel fitted with engines which may reverse to pull against the victim. The contest is unequal and the result is inevitable.

Modern guns and machinery developed by man are proving the undoing of the whale. As in the case of most other big animals, particularly where their bodies yield a profit, the whale is in danger of being wiped out—of ceasing to exist. Only proper action for its protection on the part of governments whose citizens engage in whaling can save it. That action has not as yet been taken.

Under present methods the whale, after it is killed, is lanced by members of the ship's crew. The lances they use are hollow. Through them compressed air from the ship is forced into the carcass. The lances are then withdrawn. This inflation makes it sure that the whale will stay on the surface. Having thus killed and floated one whale the vessel can leave it and pursue the herd, probably killing one or two more. It is no uncommon thing in modern whaling for a vessel to get two or three whales in a day and return to the shore station towing them all.

Here these carcasses are cut up and put to use, no part of them going to waste. Beneath the blanket of fat, which keeps the whale warm, and which is stripped off and refined for oil, is found the red meat which makes up the muscle of the huge body, a deposit of meat which may vary from five tons to many more, depending upon the size of the whale. Throughout the history of whaling this flesh until very recently has been disregarded and thrown away. Japanese whalers, however, gradually came to an appreciation of its value as human food. Now it is saved at most of the whaling stations of the world and converted into fertilizer when not otherwise used.

THE WHALE

THE WHALE 163

This meat is red, much like beef. It bears no resemblance in its nature to fish. It, of course, is not fish, but the flesh of a mammal, the flesh of an animal belonging to the same group as do the cow, the antelope, and the bear.

Because the whale is shaped like a fish and lives in the water, people have always been inclined to regard it as a fish. A more careful examination of it, however, reveals the fact that it in no way meets the definition of a fish. A fish is a scaly, cold-blooded animal, breathing through gills, laying eggs from which its young are hatched, and having flesh of a peculiar quality. The whale, on the other hand, has no scales. It is warm-blooded as are the other mammals. It provides itself with a special blanket of fat a foot thick to keep it warm. Inside it maintains a temperature normal to the higher animals.

The whale breathes air into lungs. It has no gills and cannot live, as does the fish, without coming to the surface to breathe. To be sure it can stay under water for considerable periods without coming up. Ordinarily it comes to the surface every five or ten minutes. When it does so it "blows." By this it is meant that it forces the water and the warm air from its lungs out of its breathing tubes in the top of its head. There is only a little water and much warm air. This hot air, coming into the cold of the outside atmosphere, condenses, looks like a jet of water spray, and can be seen for a great distance. Whale hunters watch for these jets of what appears to be spray. When they see them the call, "There she blows," is sounded and the pursuit is on. The whales may remain on the surface for some time or may merely take a few breaths and again submerge. They can easily stay down for twenty or thirty minutes and in an emergency for an hour. If kept down longer, however, they would drown as would any other air-breathing animal.

Whales give birth to their young as do the familiar land animals. The mother has a single young one, supposedly every two years. Newly born whales are often one fourth the size of their mothers and have been observed to be twenty-five feet long, and to weigh eight tons. This is quite in contrast to the method of the fish which lays a multitude of eggs and afterwards pays no attention to them. The mother whale manages to suckle her young just as successfully as do the domestic animals. It is interesting to conjecture on the amount of milk a mother whale produces and the amount of money that could be made from her if she could be trained to browse in the ocean and come up to the wharf for milking night and morning. The mother watches over her young one, and much affection is shown between them. If a whale hunter kills either the mother or the young one, the other cannot be driven off, but will stay about and may likewise be killed if the whaler so desires. At times whale mothers have fiercely defended their young.

The flesh of whales has none of the peculiarities of the meat of fish. It is as much unlike fish as is mutton. It is more like beef than any other meat. It is somewhat coarse and has a flavor that some regard as being strong, a fault that can be cured in the cooking. It is without bones or gristle and may be cut up into pieces conve-

AN ATTACK OF KILLERS UPON A LARGE GREENLAND WHALE, THE LATTER SHOWING INDICATION OF THE WHALEBONE STRAINER IN ITS MOUTH

nient for cooking, so that there is no waste. It has long been eaten by the Japanese and is coming into use elsewhere. Some whaling stations are now canning it.

Thus it is shown that the whale belongs undoubtedly with the mammals, with the warm-blooded animals that suckle their young. It is not a fish, but a mammal. Its flippers have upper arm, forearm, and fingers, such as no fish ever possessed. These convince the scientists that its ancestors once lived on land as four-footed beasts. They gradually became water animals and in the water took on fishlike forms because those forms enabled them better to move through the water.

Next in importance to the sperm whale among those that have teeth and that produce no baleen, are the killer whales, which are among the most ferocious of all animals. They are not important to the whale hunters, as they are only about twenty-five feet in length. In all nature, however, there is nothing more thrilling than the attack of a killer upon one of the big, whalebone whales. The killers are likely to attack in packs. The gray California whale is their favorite victim. They jam their snouts between the jaws of their huge prey, and force it to open those jaws. Then they tear out its tongue. The tongues of the big whales are huge and these killers have a special liking for them. The gray whale when attacked by killers is seized with panic, turns over on its back and makes no effort to escape. Whalers have often taken advantage of these attacks to harpoon the terrified monsters.

The familiar porpoise which plays about the harbors of the world is a lesser whale and likewise not a fish. It, too, is a strange and interesting creature with which man has had many odd contacts. Probably the most interesting of the porpoises are the so-called blackfish, which are not fish at all.

The blackfish mentally are the sheep of the sea. If one member of a school heads into a coral reef, or shoal water, or a granite cliff, for that matter, and gets a bit excited, it will rush headlong and all its fellows will follow it as though it were leading a Balaklava charge into the jaws of death while all the world looked on and applauded. And when the cohorts are withdrawn from one of these flying squadron charges and the roll is called there are usually great and yawning gaps in the lines of the school, for fishermen take advantage of the folly of the little whales and slaughter them mercilessly when they get stranded in the shoal water.

These foolish animals range from five to twenty feet in length, and are very heavy bodied for their size. The coasts of the North Atlantic serve as one of their favorite playgrounds. In the fishing grounds along Cape Cod the porpoises most frequently suffer from their frenzies. They get inside the capes and play about in the shallow inlets. Suddenly one of the big fellows of the school finds itself in water of so little depth that it touches bottom. It immediately loses its head and begins making its mad rush. If its nose is turned toward deep water, all is well and good, for it will reach safety. But if a sandy beach is straight ahead the line of action remains unchanged and soon it finds itself beached high and dry.

When the big fellow begins the disturbance the rest

of the school huddle about and begin to rush in the same direction it is taking. As an entire flock of sheep will herd together and follow the leader over a fence, so these sheep of the sea plunge headlong wherever the leader sees fit to go.

The Cape Cod villagers have many tales to tell of the stranding of the blackfish. It does not happen often, but an old fisherman will tell of half a dozen rushes of this sort in the course of his career. A man of thirty can usually remember two or three. And these recollections are occasions of great thrills and activities in the villages. Their repetition is an occasion for which all the fishing people are on the lookout, for the wide-awake may reap rich rewards when they occur.

There is a sort of a common law among the fishermen as to the division of spoils when the blackfish come ashore. The prizes belong to the men who first reach the scene and place their marks upon them. So there is always in the minds of the fishermen the possibility of the stranding of the blackfish and they are ever ready for the emergency.

The cry of "blackfish" in these villages is always as startling as was the cry of "wolf" among the old-fashioned shepherds. It means the laying aside of all other activity and a scramble for the beach and the butchery of the foolish derelicts until the sands become a shambles.

Sometimes there are only a dozen or twenty porpoises that come ashore, but the schools are likely to be much larger in size. They often run into the hundreds and there are records of twelve hundred animals that were stranded and butchered. For a village of three hundred people such an event is a great thrill and a source of profit. The foolish suicides are to be cut up and the blubber taken off and refined. Their heads must be split and the "watermelon" of oil taken out. This is spermaceti, for the blackfish is a little cousin of the sperm whale and yields an even better oil for use in lubricating fine machinery than does its big cousin. Other oil is refined from the fat of the blackfish. Its meat used to be scorned, but, with the discovery of the virtues of whale meat, it is coming to be used. The skin may be tanned, and makes good shoes and harness leather.

QUESTIONS

1. Why should the largest animal in the world be a water animal? How large is a whale compared with an elephant? How much do the largest of them weigh?
2. There are whalebone whales and toothed whales. Which is the more important? Describe the manner in which whalebone grows. What is its use to the whale?
3. What products does the whaler get from these huge mammals of the sea? Describe the peeling of a whale.
4. Did you ever read *Moby Dick, the White Whale?* It was one of the earliest American adventure stories.
5. How are the sperm whales different? What valuable store do they carry?
6. What is ambergris? How is it procured? For what is it used?
7. Whale hunting used to be a most adventurous calling. When was it in its glory? Why did it decline?
8. How was the whale harpooned and killed in the old days? Describe some of the modern developments of whaling.
9. Is whale meat good to eat? Why does it not taste like fish? To what class of animals do the whales belong?
10. Describe the points of the whale that prove it to be a mammal. Would a whale drown if kept under the water?
11. There are fierce beasts of prey among mammals on land. Are there also wolves of the sea? Describe an attack of killer whales.

12. Describe the coming of "blackfish" to a coast village. Are these porpoises really fish? What is their relationship to the whales?
13. Since man has grown powerful on the earth many of the larger animals have been killed in such numbers that they have almost ceased to exist. Only a few buffalo, grizzly bear, and American eagles are left. Has anything been done to prevent the extermination of the whales? How could action be taken to this end? Why should such action be taken?

CHAPTER XIII

THE MONKEY

S a study of animals proceeds, there is no escape from the necessity of classifying and fitting man into the niche in which he belongs. Men, apes, and monkeys are grouped together by biologists in a higher order of mammals which they call Primates. The term comes from the Latin word meaning "first." There is, therefore, a touch of the complimentary in the use of the term, although man generally may be inclined to feel a bit resentful at being put in with the monkeys. Yet here he belongs among the "firsts" of the animal world.

Biologists classify all living creatures by their physical make-up, the shape and arrangement of their bones and the peculiarities of their teeth. On this basis men and monkeys unquestionably belong together, as anybody can see by taking a look at the skeletons of each in a museum.

The bone arrangement in man and in the monkey is on the same general plan. There is the same relation between these bones as there is between those of the hoofed animals, for instance, or the rodents, or the bats. Man has taken to walking to a greater extent than have the monkeys and his legs have grown correspondingly

171

THE MONKEY IS AMUSED TO SEE
MERE MAN PLACE A LADDER TO REACH
FRUIT ON A SMALL TREE

stronger and longer, while his arms have grown shorter
and weaker. His backbone has acquired the habit of
standing up straight, while theirs has not. He started to
think a long time ago and has used his head more than
have monkeys. His brain cavity is, therefore, bigger than
that of the monkeys. Aside from these variations which
are no greater than those between one flesh-eating animal
like the cat and another like the dog, man and monkey
are biologically the same.

To man the ape and the monkey are like pictures out of

Nature's funny sheet largely because they bear a certain resemblance to himself. Man is likely to hold the ape in contempt because it has developed so little along the line in which man has developed so much. But the ape might take the view that man is a degenerate. A young Gibbon ape can leap from a tree top, hurtle through the air for a drop of forty feet, seize the tip of a branch which his weight will bend nearly to breaking, but which in the end will recoil and fling him somersaulting through the trees where he is as much at home as a puppy playing on a lawn. He too might sneer in disdain as he watched this stiff-backed, big-headed cousin of his, once like him at home in the tree tops, bringing a ladder that he might climb a few feet for a cluster of cherries.

Man has progressed greatly as a creature walking on the ground it would seem, but, from the monkey's view-point, with regard to his skill in the tree tops, he has become a third rate incompetent.

Biologically, then, man is a primate. The primates are divided into many families, one of which is the ape, another, the Old World monkey, another, the American monkey, and another, man. These other

families are no less different from man than they are from each other. With few exceptions they have five fingers on their feet, with the thumb set apart from the others that they may be fit machines for grasping. In monkeys another great toe of the hind foot is usually also a thumb, but man has been using his feet for walking so long that the great toe devotes itself to a new task—that of pushing him forward.

Most of the monkeys stayed in the tree tops and followed a different line. They changed and developed in a way that is intellectually more remarkable, but physically not nearly so great as in the case of the horse, whose ancestors likewise once lived in trees.

In their forest life monkeys are like men in many ways. In the first place they live together in groups. There is an acknowledged leader, some sturdy old male who exercises discipline and controls the actions of the others. They pick out a given place as a home and leave it only when food gets scarce, or enemies too many. They eat the food that man eats—fruit, nuts, grain, roots, onions, tender buds, and plants. Eggs are a great delicacy to them, making it hard on the birds. They are, like man, partly flesh eaters, and often feed upon insects, lizards, rodents. The crops grown by man are popular with them, such as corn, sugar-cane, apples, and for this reason they are given to hanging about settlements and to becoming a nuisance. Fences are matters of no concern to monkeys. Foodstuffs are safe neither on the ground nor in the trees and the storehouse must be locked tightly so that they cannot enter. They are frightfully wasteful and spoil ten times as much as they eat. They are treacherous, shiftless, quarrelsome, and clownish, having none of the instincts of helpfulness and service such as are shown by dogs or elephants. But apart from all this, they have a most amusing curiosity, earnestness, and prankishness so manlike in many ways as to make them an unending source of amusement. They have also an abundance of that master virtue, courage, and their devotion to their young is unsurpassed. Here again they show a manlike trait, for the mother carries her baby when danger appears and swings away with it through the tree tops.

This baby monkey is a funny looking little animal, with a face seemingly wrinkled by great age. It is, however, the chief concern of its mother, and no human parent ever lavished more care and affection on her child. To be sure, she disciplines it soundly, boxes its ears now and then, but, oddly, this is not often necessary, as the baby ape is much more obedient than the young of man. There are many known cases of mother apes that have refused to eat and have died of grief and starvation following the loss of their young ones. Their affection for men who have trained them is great and the loss of a master may lead to the death of an ape. As pets the higher order of monkeys never tire of being fondled and petted. They have a remarkable fondness for other creatures and, if, for instance, a litter of puppies is put into the cage of a colony of apes, each will adopt a puppy, will look after it with great care and affection, and will object to its being taken away later. Altogether they may be set down as most affectionate and emotional creatures, often behaving much like spoiled children.

13

The primates are mostly creatures of the tropics. Man has crowded far to the north and south, but the other families of this order live where it is warm, where there is food in the jungle the year around. In America they push as far north as middle Mexico; and on the eastern hemisphere there are monkeys in Japan. A single tribe lives among the rocks of Gibraltar, being the only monkeys in Europe.

Apes are big, tailless monkeys, most like man of all the tribes of monkeys. There are only four kinds of apes, standing at the head of which is the gorilla, styled a man-like ape and the largest of them all. The gorilla has come down out of the tree tops and is even now in a state of change, that of becoming a creature that walks on the ground on two legs. Its body is straightening up; its legs are growing longer and its arms shorter. It now sleeps on the ground with its head pillowed on its arm, as does man.

The gorilla is found in equatorial west Africa, along the coast and on the wooded hillsides of the interior. Though it can with difficulty be kept alive in climates that man likes best, it thrives in hot, humid, miasmic regions that are overrun with malaria and fever.

The gorilla is as tall as a man, but much broader of shoulder and weighs twice as much. The Natural History Museum of New York brought what it considered a typical specimen from Africa for mounting. It was five feet seven and one-half inches tall and weighed three hundred sixty pounds. Its huge chest was sixty-two inches in girth, while the spread of its arms was ninety-seven inches.

The danger of the gorilla to man, like that of many creatures of the wild, has been greatly exaggerated, as it will take to its heels if it has the chance rather than give battle. If driven to attack, however, its strength and activity make it dangerous. In a gorilla country the natives are not at all afraid of it—conclusive evidence that it does not attack man. It is a shy, good-natured creature, as are most wild animals when unmolested. It is clumsy and lumbering, does not yet often walk manlike, but rests on the knuckles of its hands. It stands up only occasionally and has the habit of beating its breast, not in rage, but in mere curiosity. Its great, hollow eyes, its hairless forehead, its ears that are remarkably human, give it a strikingly manlike appearance. In captivity it becomes very fond of its keeper, but is very exacting and insists on his constant attention, as gorillas are sociable creatures and very miserable when left alone. They do not thrive in captivity and are not likely to live long in northern latitudes, for they are fundamentally creatures of the tropics. For this reason they are seldom to be seen in circuses or zoölogical gardens. Under the care of their keepers, however, they may easily be taught to eat with spoons, to drink out of cups as do human beings, to sleep in beds, and cover themselves at night.

The chimpanzee among apes (the word meaning "little bushman" in the native language) ranks next to the gorilla in size. It walks on all fours, depending upon the calloused backs of its hands for support. It is not so far along in making the change from a tree-dwelling to a land-dwelling animal. The chimpanzee occupies a wider range through Africa and is much better known than the gorilla. It is to be found there in West Africa in a belt on each side of the equator extending deep into the interior. Its home extends right across gorilla-land. It too delights in heat and humidity. The chimpanzee is gentle and amiable, lends itself readily to being tamed, acquires good table manners, learns tricks, and can be kept alive longer in northern climes. The educated apes that become vaudeville stars are chimpanzees.

The gorilla is larger and more upright in posture than the chimpanzee, but the latter is a closer approach in structure to man when the detail of its anatomy is studied. In the detail of the number of bones in its body and the purposes they serve, it is almost identical with man. There is but one bone in the skeleton of the chimpanzee that is different from those in the skeleton of man, the sacrum, between the hips in the lower part of the spinal column. In the chimpanzee six vertebræ grow together to make the sacrum, while in man only five vertebræ are required. One may see the points at which they unite. Having used up one more of the joints of its back bone in making its sacrum, the chimpanzee has one less joint left in its spinal column. The chimpanzee and man are believed to have made themselves these different sorts of sacrums because the one suited better for the hunched position of the ape and the other worked better in the back of man who was given to standing straighter.

The orang-utan, "man of the woods" of Borneo and Sumatra, third in size among the apes, is claimed by many to be the highest type of creature of the four kinds.

It is shorter than the gorilla or chimpanzee, but has a wider spread of arms. It lives only in the dense forest of the lowlands, staying almost exclusively in the tree tops and rarely coming to the ground. It has not come out of the trees as have the gorilla and the chimpanzee. It travels through the branches as readily as men walk in paths on the ground and is said to be able to journey at the rate of five or six miles an hour, swinging from the interlocking branches of one tree to another. It is the king of the woods in which it lives, there being no animal in these islands to dispute its sway except the crocodile, a very vicious specimen of which lives here, and which it may occasionally meet when it comes down for water.

The action of the orang-utan in captivity is often quaint and curious, indicating a peculiar degree of intelligence. When it observes, for instance, that its keeper unlocks its door with a key, it is very likely to take up any piece of wood or metal that it can find, attempt to fit it into the lock, and do likewise. It is fond of amusing itself by untying knots, in performing which task it is likely to use the fingers of either its front or back feet and its teeth. If it learns the trick of untying knots, it never misses an opportunity of showing off by loosening the shoe strings of people around it.

A story of a remarkable feat of memory is told of an orang-utan which was a pet aboard a sailing ship. On that ship each Tuesday and Friday the sailors were served with tapioca, cooked with sugar and cinnamon— a dish of which this orang-utan was exceedingly fond. Soon it learned the schedule of these meals, and on those days at two o'clock it presented itself regularly at the kitchen for its share of the delicacy.

The Gibbon ape is called the alarm clock of the Malay mountaineers because of the great clatter that it sets up at sunrise and at sunset. The racket is so great that only the soundest of sleepers in all the villages in that part of the world may find his slumber unbroken after the appearance of the sun. The Gibbon is the fourth and last of the apes. It is a creature about three feet high, but much slenderer than any of its fellow apes. It is by way of being the greyhound of the ape family. It is a very shy creature of the woods and difficult to observe, but to him who sees it in action the feats of the Gibbon seem nothing less than miracles. He marvels at the manner in which it swings itself through the forest, its incomparable agility, the immense leaps of thirty or forty feet through the air from tree to tree which it seems to take, not from necessity, but for the mere pleasure and play of it.

The dog-shaped monkeys form a tribe that ranges over nearly all south Asia and Africa. They are a very populous tribe, much smaller than the apes, outstanding among which are the hanuman or sacred monkeys of the Hindus, protected by them despite the fact that they are great despoilers of crops and great thieves from their households. An odd incident in the life of these Hindus and hanuman may be seen when one neighbor sees fit to avenge himself upon another. On such an occasion he climbs upon the house of this neighbor at the beginning of the rainy season and pours rice upon the roof. This rice runs down between the tiles. The monkeys, that they may get at the rice, pull the tiles off this man's roof and throw them away. With the rainy season just coming on the household effects of this individual are likely to suffer much from the weather.

In Borneo is to be found the "long-nosed" monkey, a grotesque, hook-nosed caricature of its kind. In the forests of Abyssinia is the guereza, a creature with long, black fur tipped in white, admittedly the most beautiful of monkeys. Far flung through all this region are the guenons, the tribe from which is recruited the mass of those innumerable monkeys found in all the zoos of the world, as pets in a multitude of gardens, dancing at the end of the strings of the organ men everywhere. The guenon, raider of corn fields, is despised of all the farmers in its part of the Dark Continent.

The baboon is a branch of the primate order quite distinct from most of the monkeys. Baboons are sometimes styled "dog-faced" monkeys, and the countenances of many of them seem to bear the same relation to the face of a dog that those of other monkeys do to the face of man. The structure of the baboon is quite different from that of most monkeys. It lives on the ground, is a creature of rocks and hills, and is built for traveling on four feet. Its favorite place of abode is in the mountains of Africa and southeastern Asia. The baboon is the lowest order of monkeys and is a depraved and unlovely creature, full of cunning and malice, and given to blind rages of passion.

The American monkeys are of a family different in many ways from any of those of the eastern hemisphere. They are all small, or medium-sized, mild-natured creatures, there being no apes, no creatures comparable to the gorillas or the baboons.   They have thirty-six teeth where the Old World monkeys have but thirty-two, and this matter of teeth is considered very important by the scientists. Their thumbs in most cases are unimportant and work on the same side of the limb as do their fingers or are absent, as in the spider monkeys. Old World monkeys have long, straight noses with vertical nostrils, separated by a thin, narrow wall. The New World monkeys have flat nostrils set far apart. Because of this, American monkeys are classed as "broad-nosed monkeys." But, strangest and most individual of all qualities, they have developed the knack of using their tails as a sort of fifth hand. They can hold on with their tails. When they go to sleep, they leave their tails wound about a limb as a safety device since, if they fall off, their rope-like tails will save them. One of these monkeys can even loop its tail about a bough and swing comfortably in the breeze with no other tie to the world. The tail has a grip that is securer than that of any other member. It is highly developed and sensitive, and is often used as a feeler. The use of this tail gives American monkeys a decided advantage over those of Africa and Asia.

Some monkeys, living in trees, make themselves "nests" in which to sleep. This is done by breaking off branches and crisscrossing them in some convenient place. These nests may be deserted the next day and others built when bedtime comes. These inhabitants of the tropics have no burrows or dens, for it is nearly always warm and there is always an abundance of food. There is no need to store food or to seek protection from

THE MONKEY 183

the cold. Life for them is without some of the stern
realities that are faced by animals farther north.

Monkeys are primates. They have hands. They have
brains of a quality superior to those of other animals
except man. One member of their tribe has emerged
as the dominant individual of the animal kingdom.

What might have been the result, one may ask, if the
Egyptians had domesticated the monkey six thousand
years ago, instead of the cat, and given it the advantage
of those centuries of close association with man?

QUESTIONS

1. We have examined a number of types of animals such as fishes, amphibians, and reptiles and fitted them into their places in the animal
   kingdom. What group of animals stands highest in the animal world?
   What sets the members of this group apart from the other animals?
2. What are the leading members of the primate group? What traits
   do they possess that are manlike?
3. Apes are the most highly developed of the primates next to man.
   Name the four kinds of apes. Describe gorilla life. Do you see
   signs of progress toward man in the gorilla? What manlike
   tricks do they learn in captivity?
4. Have you ever seen a gorilla? A chimpanzee? What is the one
   important structural difference between a chimpanzee and man?
5. What are the other two apes? Describe some of the tricks of
   these animals.
6. Tell the story of the ape that came to the edge of the woods.
7. What are the hanuman? Where does the monkey of the organ
   grinder come from?
8. The unloveliest of all the primates is the baboon. How does it
   differ from its cousins?
9. There are some most peculiar differences between American and
   old-world monkeys. Point out some of those differences. The
   differences indicate that these families have been separated for a
   long time and have developed along different lines. Can you
   point out widely separated races of human beings that have
   grown always more different?

10. What surprising use does the American monkey make of its tail that his eastern cousin has not learned? Do you suppose the American monkey has learned this as a new trick or is it a thing that all monkeys knew at some time, but which old-world monkeys have forgotten?

11. In all of this what do you find that proves the primates to be of a higher order than the other animals?

CHAPTER XIV

THE CORAL POLYP

OST ocean bathers have been stung by sea nettles. When this happens, a certain hungry huntsman of the deep has launched a thousand javelins, each with a string tied to it, and every one of them has gone home. This huntsman has expected to drag food into its yawning maw and make a square meal of it, or maybe it has sought, in this case, to drive away a prowler that it recognized as being too large for it to lead to the slaughter. One cannot be entirely sure of what is back of the sea nettle's action.

This sea nettle contact will serve as an introduction into a curious animal society that one, ordinarily, is not likely to learn much about. Judging from the sea nettle, one would conclude that here is a most unsubstantial group. If he should put one of these sea nettles in the sun until it had dried out he would think he had proved this to be true. There would be little left but a stain.

But the sea nettle has a cousin, the coral polyp, which is made of sterner stuff. Dying, it has left to the world most substantial souvenirs of its having lived. It has left, among other things, the Bermuda Islands out in the Atlantic six hundred miles due east of North Carolina; it has left the Florida Keys and many other trimmings to the chief peninsula of the United States; it has left the Fiji Islands away down in the South Pacific; Oahu, in the Hawaiian group; the New Hebrides, and hundreds of palm-fringed islands and atolls stretching half across the world from New Zealand to Japan; it has left the Great Barrier Reef which combs the frothing waves for twelve hundred fifty miles along the coast of Australia.

The coral polyp, cousin to this frail sea nettle, is one of the greatest masons of the animal world; it has left behind it such monuments as no other animal, not even man himself, can ever hope to rival.

Yet this coral polyp, considered without the skeleton which it makes for itself, is almost like the sea nettle. It is little more than a film of flesh that holds together a group of quite primary animal organs.

But these organs are able to do one small thing upon which their greatness is founded. They have mastered one knack in building, have perfected one process of getting the material for it, just as man has learned to separate iron from red earth, and the result has changed the geography of the world. Coral polyps have learned to separate carbonate of lime from the waters of the ocean and to build bones of it for their own peculiar uses. These bones, piled up through the ages, have been a patent factor in changing the form of continents.

The sea nettle, not having acquired this knack, dies and leaves no trace. Its cousin, quite unconsciously and working only to serve its own ends, learned to separate lime from sea water, with the result that vast areas of the tropic seas of the world are studded with palm-fringed islands which become the happy abiding places of plant and animal life and even of man himself.

The animals of this group are called polyps because they have many feet. A polyp is literally a many-footed animal. It has many feet if one may consider the divisions of its body which grope about for food as feet. It is the flower animal, growing there in the ocean. The

LARVA EMERGING FROM THE MOUTH OF THE MOST COMMONLY KNOWN RED CORAL

sea anemone, as radiant and colorful as any in your garden, is a polyp. The coral polyp is a flower-animal, a sea anemone, which has learned to take lime out of the water and build a skeleton of it.

It is interesting to trace these odd flower-animals from the cradle to the grave and note the peculiar experiences through which they pass. A polyp begins as an egg about the size of a pinhead. Its mother releases it there in the great ocean stretches and never sees it again. It floats or swims about for a few days, possibly for a week or two. It is a free thing, unattached. This is its only span of freedom, the only period of its life when it has a chance to travel. It may get into some ocean current as, for instance, the great Gulf Stream, and ride far away and, if ambitious, may start itself a new reef which, in a million years, may grow into a new island paradise for which a mark is made on the map.

The egg becomes a larva, a baby polyp, with tiny bristles all over it which it uses in swimming, later becoming pear-shaped or cup-shaped. It is now ready to settle down. It selects a nice hard rock or a big mollusk shell, glues itself to it, and is attached for life.

In the meantime it has started a dimple in its upper end, the end that is to become its mouth. The coral polyp now develops a stocky stem with a flower-like head. The dimple of a mouth opens up; the inside takes form; organs develop. Chief among these is a stomach which must be fed. The flower-like head begins to put out tentacles, or arms, often brilliantly colored, which look like the petals of a flower. Early observers, judging by appearances, thought these polyps were, in reality, plants.

These tentacles, or petals, in fact, the whole surface of the animal, are abundantly supplied with weapons which take the place of spears or harpoons used in hunting or defense. These are very small and are hidden away in microscopic cells which some students call lasso cells. In each of these cells is kept a long string or lasso,

coiled up like so much rope. When the lasso is thrown
the string is turned inside out as might be a stocking with
a weight in its toe. The end of the lasso is sharp and may
serve as a sting. If the shot is successful the game is
harpooned. So penetrating are these tiny darts that they
have been found driven entirely through the two hard
shells of tiny mollusks. No wonder a thousand of them
nettle the flesh of a bather. Less is the wonder when
it is understood that each lasso point carries poison
intended to quiet the harpooned game, the victim as
big as a pin point, wriggling there on the end of this
tiny line.

Just as man goes harpooning for whales and shoots
his darts from cannon, so does this polyp shoot its lances
from microscopic cells into the game for which it lies
in wait. Its game is often animals which are so small
that they cannot be seen with the unaided eye; again,
it may be water fleas which may be plainly observed.
It is mostly small crustaceans which are eaten shells
and all, but it may be crab meat which man feeds it by
hand.

The tentacles, covered with lasso cells, cluster about
the mouth of the polyp. They catch and sort the food,
work that which is acceptable toward the mouth, comb
out that which is not wanted and keep pushing it away.
In this work they show plainly that the polyp is chiefly a
carnivorous animal, a flesh eater.

After the polyp has attached itself—has taken form
and developed organs—it begins that process which makes
of it the master mason, the separation of lime from the
ocean water. This task it performs quite unconsciously,

of course, just as our digestive organs select from our food the materials that go to make our bones.

The water of the ocean is a great storehouse for lime just as it is a great storehouse for salt. It is easier to tell that the salt is there because it tastes so strongly, but all ocean water carries much lime as well as salt. Most of the streams that run down from the land, as, for instance, those of the Mississippi Valley, carry great quantities of lime taken from the limestone deposits over which they have passed. They carry this down to the sea just as they carry salt. The air evaporates water from the ocean which the winds carry back to the land where it becomes clouds and rain. Only pure water comes back, and so the lime and salt are left behind and the ocean gains more lime and salt all the time.

Before the rivers get to the ocean with their lime, however, certain animals have begun taking it out and putting it to their peculiar uses. The fresh water mussel, for instance, has taken out lime for making its shell and has thus furnished mussel shells for the manufacture of pearl buttons. The oyster takes the lime it needs from the water for building its two convenient half-shells. There in the ocean mollusks of uncounted varieties take lime from the water for similar purposes. Tiny, microscopic animals make themselves shells and die. These shells fall like rain to the floor of the ocean where they slowly build up lime deposits. Oysters and other mollusks leave shells which help to form layers of limestone hundreds of feet in thickness. They, too, are land builders. All work together in turning this lime, which the rivers brought from the land, back into land. Now comes

the coral polyp, glued to its resting place, which likewise has solved the secret of lime separation from water
and which has become the master builder of them all.
Its results can be more easily appreciated than those of
the others because it concentrates its building at given
points and pushes its structures to the surface.

This typical coral polyp, with its thick stem and its
flower-like head, begins to gather lime and store it away
in its body for purposes of stiffening and protecting that
body. With a rugged, mineral frame back of it, it can
the better stand out boldly in the breakers where the
food supply is richest and attend to the all important
detail of keeping its bread basket full. Into this mineral framework it can withdraw the softer portions of
its body for safety when there is danger abroad. It is
for these two purposes that it builds its rugged framework.

This typical polyp might have a body a quarter of an
inch thick and might be two inches long. Such a polyp, it
should be understood, is but one of thousands of varieties,
each of which has the same fundamental structure, but each
of which has a different form from the others. While the
polyp we have described stands out quite prominently,
there are other kinds that cling to the rocks like moss
flattened upon its surface. Still others are packed together like tubes of macaroni so arranged as to be called
pipe organ coral. Others radiate around the center as
might a many pointed star, sometimes forming figures
not unlike enlarged snow crystals. Yet others build
structures with widely separated prongs that look not
unlike the antlers of a deer. There are forms in which

14

a single polyp builds a shell as big as one's two fists with many ruffles running out from the center. There are forms in which colonies group themselves into ovals with many wrinkles, appearing much as would the exposed brain of the human being and, therefore, called brain corals. There are colonies that grow in huge clusters as big as hogsheads and so add much to the reef of which they are part. There are numberless forms that all tie back to the principle of the typical polyp—all lead the life that it leads and complete the cycle which it follows.

Aside from this cycle of life from the egg to the reef builder, the polyp has another method of increasing its numbers. When it is planted on its rock it produces new individuals by a process known as budding. New individuals develop and separate themselves from the parent body and thus become independent polyps. They grow much as might a shrub branching at the roots. In this way a single polyp becomes a colony. In this way the cluster that is known as a brain colony may develop into thousands of individuals all having their origin in the one parent polyp. In this way this one tiny parent

PIPE ORGAN CORAL

polyp may add to the coral reef a bulk of stone as big as a football or sometimes as big as a barrel. In this way staghorn corals may grow into a forest of branches that beat the breakers into a froth.

Coral polyps thrive only where the water is warm. Most coral-building polyps grow between the parallels of latitude 30 degrees north and 30 degrees south of the equator. It is a peculiar fact that coral thrives only on the western shores of oceans. The great ocean colonies of the Atlantic are along its western shores, in the Caribbean Sea and the ocean itself as far north as the Bermudas, which form the important coral colony farthest from the equator. These colonies, it will be noted, are in the course of the great Gulf Stream which brings warm water from the region of the equator.

In the Pacific Ocean, likewise, the coral colonies are in the western part of that ocean. They extend up past the Philippines as far as China and south from the equator to and beyond the Australian shore. There in the Pacific, where they abound, warm ocean currents are to be found that carry abundant supplies of food trailing over their door sills. The great ocean currents which develop, flowing north and south from the equator in both the oceans, are on the west fringes of those oceans, a fact which is due to the rotation of the world on its axis which piles the water up on the west sides of oceans. They carry warm water north and south. On the east sides of oceans the water is colder, since it flows down from the north.

Most coral polyps live in comparatively shallow water, water that ranges from the surface to a depth of not more than one hundred feet. It is these surface waters that abound in the food supplies that sustain coral polyps. Thus it comes about that reef-building corals live in the shallows of warm, moving waters in the west sides of the oceans.

To be sure there are certain corals that do not do just this. There is, for instance, the sort of beautifully colored coral from which jewelry is made. Most of it comes from the Mediterranean Sea, on the coast of Italy and France. The polyps that produce it make no reefs. Their nature is such that they do not pile up at particular spots. They scatter about. Coral fishermen go out and drag for them.

A natural place for reef-building coral polyps to start to grow is along the shores of a tropical island where the water is warm and clear and the bottom is rocky. Along such shores the floor of the ocean is likely to be covered with a growth of coral as a meadow is covered with clover.

The growth of these coral beds is very slow. The coral itself may grow half an inch a year or, in the Pacific, as much as two inches a year, but the reef as a whole grows more slowly. If one of them should approach one foot nearer the surface in the passing of a century it would be doing fairly well. As islands grow, however, centuries are very short periods of time, and a coral bed twenty or thirty feet below the surface would steadily push upward until, in the end, it would arrive there. As it came close to the top of the water, a new set of conditions would develop that would cause it to grow more rapidly on the outside than near the shore. Near the

shore sediment might wash down and fragments from the
outer part of the reef might drift in. These would inter-
fere with the growth of the reef. The food supply closer
to shore might also be less abundant than it was on the
outside of the reef. It would, therefore, be likely to happen that the reef would grow more rapidly on the outside than on the inside. Thus would the coral reach the
surface first at a little distance from the shore. So
would it create what is known as a fringing reef, a reef
that runs along the shore a few hundred feet from it,
with a lagoon of comparatively shallow water between
it and the land. These fringing reefs are found more
often than any other form of coral structure.

Another type of land which the corals build is the
atoll. Atolls are reefs which extend in rough circles inclosing bodies of still water. They protect these bodies
of water from the ordinary wave action of the outside
ocean. They create those peaceful havens in the midst
of great reaches of restless ocean found here and there
in the Southern Pacific.

There is much speculation as to how atolls are formed.
The most popular theory is that they began by being
fringing reefs to islands that afterwards disappeared.
In some parts of the world the surface of the earth is
slowly sinking. Under such circumstances the corals
might start building a fringing reef around a rocky island.
That island might slowly sink into the sea, but the corals
would keep on building and would keep the reef near
the surface after the island had disappeared. So an atoll
would come into being.

Yet another theory is that the fringing reef encircled

an island which was worn away by the action of the wind and waves and in the end disappeared beneath the sea.

Probably the largest single structure in the world built by corals is the Great Barrier Reef which runs for twelve hundred fifty miles along the northeast shore of Australia. This is the largest and longest of all reefs and, oddly, it is at places as much as ninety miles from the shore. It is a great frothing area of many forms of growing coral coming barely to the surface of the water. It incloses a lagoon of comparatively shallow water ranging from ten to ninety miles in width along the coast of Australia.

This type of reef is called a barrier reef. The theory of the existence of barrier reefs so far from shore is that they began as fringing reefs close to the coast; that the shore line slowly submerged; that the reefs grew as rapidly as the coast line sank; that they grew more rapidly on the outer side than on the inner side, and, consequently, became always farther and farther from the receding shore. So, today, this Great Barrier Reef survives, far out from the Australian shore, as the greatest coral wilderness of them all.

When the coral has reached the water surface either as a fringing reef, a barrier reef, or an atoll, there remains the further step that converts it into an island on which vegetation will grow and upon which even human beings may come to live. The waves beat on this coral reef, break fragments from it, and throw them on top of it. Some of these fragments get into crevasses where they lodge. Some of them are ground into sand which fits itself into smaller openings. There come to be masses of snow-white coral sand around most reefs and this may be used by the waves in island building. The mollusks are always present in great numbers in these waters. They die and leave their lime-built shells as a contribution to the structure which is given form by the coral polyps. These polyps, as a matter of fact, furnish a minor part of the material that goes into these reefs, but they set up the framework which holds together material from many sources.

The waves beating on the reef carry seaweed which drifts in and is entangled among the prongs of the coral. Streams from the mainland may bring down mud or driftwood. The carcass of a whale, giant of the deep, or the hulk of a wrecked merchantman, silent evidence of some tragedy of the sea, may make their contributions toward land building. All unite in a formation that in the end provide sufficient soil to support vegetation.

About this time there appears floating on the surface of the ocean a peculiar object that looks not unlike a cigar. It is in reality the seed of a mangrove tree that has fitted itself out in this way for the very purpose of riding the high seas in search of a new home. It floats nose up, stern dragging heavily in the water. In this position it is cast easily high up on any beach that it may accidentally encounter. On the heavy end are threadlike roots, and these sink quickly into the sand. The action of the waves beats out a groove for the body of this cigar-like seed. It becomes fixed. From it springs a mangrove tree that, before long, may be thirty feet high. Its roots, much of them above ground, form a net for catching drift. Its branches droop to the ground and again take root. The mangrove tree becomes a trap for drift, and helps the island to grow. Other seeds are added by chance and soon there is luxuriant vegetation.

As we have said, the Bermuda Islands, far out in the Atlantic, are coral built, and are farther from the equator than any other group of such origin in the world. This may be explained by the fact that they are in the midst of the great Gulf Stream which brings warm water from the tropics. Baby coral polyps, before they settled down, must some time have taken a long ride from the South in this warm Gulf Stream to have planted a colony here.

Off the southern extremity of Florida, as everybody knows, is a flock of low-lying islands, the Florida Keys, many of them inhabited, with the building of which the coral polyp has had much to do. Much of a great area there is slowly approaching the surface. The Keys are the portions of it that have already reached it. Among these growing Keys and along the mainland the water has become very shallow. It is when water gets shallow that settling basins are formed which gradually fill in and build more land. Flats are formed first which develop into low-lying land like that of the rest of southern Florida.

Out beyond the Florida Keys is another reef with just an occasional sand bar bearing a beacon as the first evidence of its approach to the surface. As time passes another series of keys will grow out there, will fill in their gaps, and will weld themselves into a new outrider for the southern end of the Florida peninsula. This, in turn, will be followed by another rim of keys. So will the reach of what is now the United States be pushed always to the southward.

This growth of a continent, however, is very slow. Ten thousand years is a time so short that it brings about little change. Ten thousand years in the story of man, however, is a time so great that history may not span it. These corals cannot be hurried.

QUESTIONS

1. What animal is the master mason of the world? What structures have you seen that it has built? What is the secret of its success? Do you know any of its cousins?
2. The polyps are animals that look like flowers. Show how their life cycle proves them to be animals and not plants. How does the polyp settle down? How does it snare its food?
3. Where does the coral polyp get the lime for its shell? Innumerable creatures build themselves shells of this lime, die, and the shells fall to the ocean bottom. How does the coral polyp's method of building differ from this?
4. Describe some of the different forms of coral. How do the polyps multiply? How do they start a reef?
5. Why do we find coral islands in warm, flowing water? In what ocean is the mass of the coral-built land of the world?
6. The polyps that make the colored coral used in making jewelry are a different breed. Where do they come from? Do they build islands?
7. Coral polyps thrive in shallow water because food is more abundant. Show how a fringing reef is built up. How are atolls formed?
8. Describe the Great Barrier Reef of Australia. How is it believed to have come into existence?
9. How does a reef become an island? How is vegetable matter contributed?
10. Describe the coming of the mangrove tree. How does this tree help the island?
11. How did the corals happen to build the Bermuda Islands so far from the equator?
12. Where are the corals adding to the land of the United States? Describe the manner in which row after row of keys are being created.
13. Are the corals vertebrates or invertebrates? To what class do they belong—mollusks, vernus, arthropoda, or what?

CHAPTER XV

THE OPOSSUM

HE opossum is the granddaddy of all the more advanced American animals. It is the oldest of the order of mammals, the milk givers, that survives in this country. It is of an order which had great prominence in the world before such present-day, ordinary kinds of animals as bear, deer, rabbit, monkey, man appeared. It is a hangover from the early ages when most of the world was inhabited by a lower order of animals—an order so stupid and inefficient that nearly everywhere it has ceased to exist. The opossum is a remnant of a race surviving, strangely, among a host of other and more intelligent forms of animal life which have crowded its near relatives from most of the earth. The opossum lives only in North and South America, and its surviving relatives are to be found nowhere but in Australia.

Being of an old and almost extinct order, the opossum may be expected to have some odd features which are unknown to the other animals with which we are familiar. It has, in fact, a number of such peculiarities, but the strangest of them all is the pouch which it possesses on its under side, and in which it carries its young.

200

Only mother cockroach, oldest of the insects, with her satchel for her young, has anything to compare with it. It is most unusual that an animal should have a pocket in its skin just as human beings have them in their coats, but this is true of the opossum. All the animals in the world that have these pouches for their young are relatives of the opossum, and they are all put in a class together—a class known as pouched animals or, as the scientists would say, "marsupials," the word being a form of "marsupium" which means, in fact, a pouch.

When the marsupials developed pockets they had a very special use for them. They were not to carry such odds and ends as schoolboys prize, but were to contain one thing—their young ones. Only the female marsupials have pouches and they use them only when they have offspring of which to take care.

The opossum is the only pouched animal that exists anywhere in the world outside of Australia. There was a time when such animals were to be found in abundance over most of the earth. Fossil remains prove this, such remains being particularly abundant in England, the other side of the world from Australia where pouched animals now abound. As time passed, however, the world developed other varieties of animals, such as the flesh eaters of today, the cats, wolves, and bears, and they gradually killed off the slow and stupid marsupial. In all Europe, Asia, and Africa there is to be found today no representative of these animals which once overran the world. In America there is only the opossum which has managed to survive in favorable regions, from the eastern United States far into South America. Here in a fertile region where food is abundant, where forests furnish favorable hiding places, and where its enemies have not been too plentiful, the opossum has been able, through the centuries, to keep alive.

With this gradual disappearance of the pouched mammals from the rest of the world it is a curious fact that they survive abundantly in Australia and some near-by islands. There, in fact, they make up the great mass of the mammal life. Aside from the dingo, or wild Australian dog, and a rat, there is no native mammal in Australia which is not provided with a pouch.

When the ships of western nations first began going to Australia they marveled much at the strangeness of the animals they found there. It was the famous Captain Cooke, English voyager, who, a century and a half ago, is alleged to have been eaten by cannibals of the South Seas, who first reported to the western world the existence in Australia of these queer, pouched animals. Captain Cooke first described that giant of the pouched animals, the kangaroo. It was he who gave it this odd-sounding name, supposedly getting it from Australian natives. Later students, however, have not been able to find this word in the Australian dialects, so its real origin is a mystery.

The fact that the animal life of Australia is so different from that of the rest of the world is based on geology—on that phase of geology which has to do with the movement of the earth's crust up and down. There was a time when Australia was connected with the mainland of Asia, and that maze of islands which now exists between it and Asia was then the highlands of a great stretch of country which has since sunk beneath the sea.

Australia was cut off from the rest of the world at the time when the pouched animals were everywhere plentiful. The types of animals that afterwards developed on the other continents, and which gradually replaced the clumsy and stupid pouched animals, did not appear in Australia. There pouched animals, in the absence of these fitter rivals, continued to be very abundant. As the ages passed they developed into different species, each of which, under the influence of differences of conditions, showed distinctive peculiarities.

There was, for instance, the kangaroo, the biggest of all marsupials, which lived in herds on the open plains. The great plains of America, starting with a quite different animal which once lived in swamps that covered that region, developed the buffalo. The great plains of Australia developed as its outstanding inhabitant this odd creature, the kangaroo.

The largest of the kangaroos weigh as much as a man. When they stand upright on their hind legs for a look about them they are as tall as a man. When they settle down into a more comfortable position, that in which they rest upon their two hind legs and their massive tail, forming the third member of a tripod, they are about four feet tall.

When the kangaroo took to the plains it chose leaping as its method of travel. It stood upon its hind legs and jumped. When it did so it did not alight on its front legs, as does the rabbit, but again upon its hind legs. The gait was rapid, springy, and not ungraceful. The animal did not use its front feet at all when running. The front feet and legs, therefore, became small, while the hind feet became very powerful. The tail, which is used as a prop when the animal is still and as a balancing pole when it is jumping, is a thick, powerful organ tapering to a tip. When nibbling grass the kangaroo rests on its front feet and upon this tail, swinging its great hind legs as though it were supported on crutches. When it is pursued by dogs and goes leaping across the plain it covers fifteen or twenty feet at a jump and makes speed equal to that of a running horse.

The kangaroo is a timid animal, but when cornered by dogs, with which it is hunted, it may turn upon them, place its back against a tree, and fight. Then the single great claw on one of its toes becomes a dangerous weapon and may rip a dog open at a single stroke, or may disable a man.

The mother kangaroo gives birth to but a single young one at a time, which is surprisingly small and helpless when born; not as big as a mouse. The mother promptly puts the baby into her pouch, where it attaches itself to a teat and grows rapidly, and when it is eight or nine months old it weighs from ten to fifteen pounds. At first it is blind, hairless, and helpless, with its limbs and other organs only slightly developed, but it grows rapidly and soon may be seen putting its head out of the pouch and looking about as its mother travels jerkily over the plains. A little later it comes out and plays about, but scurries hurriedly back to its hiding place upon the appearance of danger.

Hunters of kangaroos have reported that a most peculiar thing sometimes takes place when a female kangaroo with a heavy young one in her pouch is being pursued by dogs. The weight of the young one, naturally, is quite a handicap for the mother to carry. If the dogs get too near, if it appears that they are about to catch her, the mother will reach into her pouch, take the young one out, and fling it as far as she can to one side, and thus lightened, she may escape and later return and find her baby.

The wallaby is a smaller type of kangaroo, weighing about sixty pounds. Both wallabies and kangaroos are good to eat and the skins of both furnish leather that is excellent for making shoes. There are other kinds of marsupials in Australia which have the form and habits of squirrels and live in the forest. Another type has a form and habits somewhat like those of a mole. One, the wombat, is a short-legged, thick-set fellow, as big as a small dog, living in the forests, which shuffles about in a way which suggests a bear. In Tasmania, the island to the south of Australia, there is a ferocious, flesh-eating, zebra-striped marsupial, known as the Tasmanian wolf. Thus have these pouched animals, living in their isolated corner of the world, developed forms counterfeiting the flesh eaters, the rodents, and some other types of mammals found elsewhere.

The opossum, which is the only survivor of the pouched animals outside of Australia, strangely, does not exist at all in that continent. Its species has been so long separated from the original stock that it has grown quite a distance away from it. It still retains, however, the peculiar traits that make the marsupials different from any other order of mammals in the world. Like them it has a different brain formation, the two hemispheres being much more distinctly separated than in any other mammal.   Like them, it has the pouch for its young.   Opossums are virtually without vocal organs, as are all the marsupials.   None of them appears to have any effective noise-making organs.   They are capable of little more than a sort of heavy breathing or hissing which can be described as a barking only with a stretch

THE OPOSSUM'S USEFUL TAIL

of the imagination.   They do not have lids to their eyes, but merely a sort of membrane with which they cover them.   They are stupid and helpless in a way peculiar to the pouched animals, and are yet the possessors of a cunning which helps them to survive.   The opossum is a gray, scraggly haired little beast, with beady eyes like a rat, a snout like a pig, pink skin like a white mouse, and a firm, hairless tail, which it can coil about the branches of trees so that it can swing at ease with its feet free for gathering fruit.

The opossum is a woods animal. It lives in the forests, usually making a hole among the roots of a tree or using a natural hollow in the tree for its nest. Many a hunter has found that it is particularly difficult to dig among these roots to get his opossum. The opossum can climb trees with great ease, can run along their limbs, and pass from one to the other. It likes to curl up on a branch well toward the top of the tree and sleep there, safe from the intrusion of most of its enemies. It is, in fact, quite at home in the tree tops, more so than is our familiar friend the cat. Its widely separated toes enable it to use its forepaws to grasp objects as with quite human-like hands.

One of the facts which has helped the opossum to survive is the readiness with which it eats almost anything which comes its way. It eats in quantities, for it has a huge appetite. Whenever opportunity occurs the opossum is a flesh-eating animal and preys upon mice and rabbits. It is given to invading chicken roosts and is a demon for slaughter. It is an insect-eating animal and has a special fondness for fat grubs which it digs out of rotten logs. It likes eggs and is a robber of birds' nests as well as a thief of eggs from the chicken house. Apples, carrots and onions, and fruits of many kinds are delicacies for its bill of fare.

The time of great feasting for the opossum comes, however, when grapes and persimmons arc ripe in the fall. It is then a fruit eater, and feeds so greedily, almost to bursting, hanging there by its tail, that it becomes so fat it can scarcely waddle.

15

In November the opossum becomes the great prize of the huntsman, who regards it highly as an item for food. Then opossum hunting in the Middle and Southern States is considered first-class sport and results in the capture of prime fat fellows, suitable for the oven. It is in November that parties are most likely to go out by night, carrying lanterns and axes, and accompanied by dogs with a special training for this type of sport. The dogs locate the opossum and chase it up a tree, which is then chopped down, that the fat gormand may be captured.

A large opossum may weigh a dozen pounds, and when it is in the right condition is regarded as a great delicacy. It is something like a suckling pig—fat, tender, and savory. It is best when baked and fringed by sweet potatoes.

It is upon such an occasion as its capture that one of the peculiarities of the opossum is sure to show itself. When surprised while robbing the chicken house or when otherwise brought to close quarters, the clumsy opossum, which is almost defenseless and is a slow runner, resorts to an attempt at deception which is all its own. It pretends that it is dead. It "plays 'possum." The uninitiated, if there are any such, might believe that the opossum is actually dead. Its eyes are covered with a film, it is limp and motionless. It makes no response to any degree of torture that may be inflicted upon it. It maintains its lifeless appearance in the presence of every attempt to arouse it—every attempt but one. If this seemingly dead opossum is thrown into the water it quickly comes to life and starts swimming quite ably to land.

This attempt on the part of the opossum to deceive its captor, be that captor modern man or beast, is almost never effective. So useless is it that students of animals have marveled that a product of nature should ever use it. No hunter of the opossum, man or beast, is likely to lay it down unwatched, thus giving it a chance to sneak away.

One theory as to why the opossum pretends to be dead is that at some time in its past its chief enemy was an animal which preyed only upon others which were alive, and had no interest in an animal which was dead. The toad, for example, catches only moving insects, and there are caterpillars that roll up in balls and remain still to disarm it. By seeming to be dead in that past age the opossum might have escaped this enemy. Playing dead may be a trick that has come down with it from a time in the past when it was useful.

A characteristic of the opossum which has done much toward keeping it alive among competing mammals more active and more intelligent than itself is the great number of young it bears. The female opossum may have three litters of young ones in a season and each litter may contain from eight to twelve. Thus does she bring into existence such a number of her kind that the great majority may perish and the race will still be carried on.

As in the case of the kangaroo, the opossum puts her little ones in her pouch when they are born. They are very small and undeveloped at birth and grow slowly in the pouch, but in time they come out and begin to play about, scurrying back again at the appearance of danger. They use the pouch until they have become so large they can no longer get into it. Even then they continue to stay close to their mother, cling to her hair and in some of the species resort to a very odd method of riding on her back. The mother of this species turns her tail over her back like a trolley and holds it there. The young ones climb upon the mother's back and hold to her fur with their feet, at the same time wrapping their tails around hers, thus giving themselves an additional firm anchorage. There, a dozen in a row, they give an odd exhibition of bareback riding.

Altogether, they are a stupid, unprogressive race, these marsupials, the lowest order among all the mammals, as poorly fitted to survive as is the oxcart in the age of automobiles. But they are interesting, largely as a relic of the past, just as Pompeii or the ruined temples of Yucatan are interesting. Their day is gone. Modern competition in the animal world is too strong for them. They are passing.

QUESTIONS

1. What is a marsupial? What marsupials are to be found in the United States? Where are marsupials abundant?
2. What can you say about the age of the marsupials? Are they a disappearing race? Why? By what means have the opossums managed to keep alive in America?
3. What are the outstanding peculiarities of the marsupials? How do you know that they are mammals? How do they use their pouches?
4. Aside from the dingo dog and the rat, all the native mammals of Australia were marsupials. Even this dog and rat may have been introduced by man. How do you account for the fact that these backward animals still hold sway in that part of the world?
5. The kangaroo is the biggest of the marsupials. Describe its life on the Australian plains. How does it run? How does it use its tail? Tell of the care of the young kangaroo.

6. What other marsupials are found in Australia? Do you think it might be possible to domesticate the wombat and use it as a meat-producing animal?

7. How are an opossum's brain, its eyelids, its vocal organs, its tail, different from those of other mammals? How do you account for this?

8. What is an "arboreal" animal? How do his arboreal habits help the opossum to survive?

9. What does the opossum eat? How is it hunted? How does it try to deceive its captors? How can one test an opossum to find out if it is dead?

10. One of the laws of Nature is that its creatures shall bear young in proportion to the dangers they face. If many are born, some may escape. How many young opossums are born in a single season of a single pair of parents? What does this indicate? How does it compare with the number of young the kangaroo bears in Australia? What does this signify?

Chapter XVI

THE FUR SEAL

HERE on the Pribilof Islands there is a colony of Uncle Sam that is one of the most unique settlements under the sun. Here a group of individuals nearly a million strong has its home on two islands a dozen miles long and on three jagged rocks, jutting out of Bering Sea far the other side of that long arm of Alaska which reaches out toward Japan. There are endless ice drifts to the north. Over to the right, two hundred miles away, is the waste of the north Alaskan mainland. Siberia is yonder to the left three times as far. As a crow flies it is two thousand miles to Seattle, which is the United States. All around lie the fog-covered stretches of Bering Sea on which the sun rarely shines.

The individuals that go to make up this colony are not human beings, though in its organization they lay down a scheme which is not unlike that of a human settlement. They are members of a queer race nearly all of whom in the whole world live right here—a useful race which Uncle Sam has saved from being sponged out of existence. There would have been none of them alive today but for Uncle Sam. They are, in fact, the fur seals—those seals which furnish the choice skin for milady's coat the world around.

Ninety per cent of the fur seals of all the world live upon these small islands. A thousand miles away, on the Russian side of Bering Sea, are the Commander Islands, which have a small seal colony, as has Robben Island belonging to Japan. There is a small fur seal colony in Uruguay and others in the Antarctic Ocean and a mere family here and there on the Pacific coast of South America. Once there were millions of fur seals of the tribes that came from the far South where again it is cold, but there is the barest remnant of them left. This American colony is now ten times as numerous as are all the others combined.

All of this, of course, is exclusive of the hair seals which inhabit much of the ocean space of the world. Their skins are valueless, their chief usefulness lying in the oils that are refined from them and their ability to balance balls on their noses in circuses.

The history of fur seals is an interesting one. A hundred years ago they roamed freely over all the Pacific. There were great rookeries of them in the north and they swarmed south to Japan, to San Francisco Bay, where they were very abundant, and to the northern shores of Lower California. They came up in the same way from the Antarctic. The southern fur seals had rookeries on the coast of Chile and Peru. Great areas of both the north and south Pacific abounded in these animals which were clad in skins that dressed softer than velvet and wore like leather.

To the Pacific came the early fur hunters and these simple-minded seals on their rookeries or breeding places fell easy victims to them. It was only necessary for the hunters to visit the rookeries during the summer season, the breeding season, when all seals go ashore, and slaughter them at will. There on the rocks the seals were helpless and unresisting victims. Beasts of prey might have eaten their full of them and gone away to sleep in stupid satisfaction, but these man creatures never got enough of slaughter.

So it happened that almost all the fur seals of all the world were slaughtered. There were colonies of them left only in obscure nooks lost in the far North, at distances so great that hunters did not often reach them. Chief among these, as the last century drew to a close, was the colony at the Pribilof Islands. Here, back of the Aleutian chain, where there is so rarely any sunshine to torture them, there existed, no longer ago than 1880, herds of seals, millions strong. Here, the United States, after she bought Alaska from the Russians, farmed out the privilege of taking skins; and here, for decades, it received each year tribute on one hundred thousand of them.

Then it developed that fur seal hunters from many nations found there were profits in killing these animals, and they came into these waters to engage in what is called pelagic sealing, taking them on the high seas. They killed the seals as they swam back and forth to the breeding grounds, mother seals whose pups would starve, young seals not yet mature, seals whose bodies sank and were not recovered after being killed. They killed without control, wastefully. It became plain that even the fur seals of this last retreat were soon to be wiped out.

By 1910 the seals of the Pribilof Islands, the biggest

herd in the world, had been reduced in numbers to a little more than one hundred thousand. It was in 1911 that a treaty was signed by the United States, Great Britain, Russia, and Japan, providing for the protection of the remaining seals and making it unlawful to hunt them at sea. Thus, by a single act, was the northern fur seal saved to the world.

The United States owned the Pribilof Islands, the greatest of the seal colonies. Since 1912 it has carefully looked out for the welfare of those islands and built up the numbers in the herds. There were some two hundred thousand seals that came to the islands in 1912, and ten years later that number had increased to six hundred thousand. So was it demonstrated that under proper care this herd might be built up to number a million, even two or three or four millions, in the course of a few decades, if the Government should see fit to develop it, should consider such development wise. In this way the United States Government came to control directly the main fur seal skin production of the world and to own one of the most remarkable groups of animals in existence.

It is the choice of the seals of the Pribilof Islands that this should be their home. It is their fixed habit that they should refuse to colonize anywhere else. They roam over a vast feeding ground, spread out over much of the north Pacific and, in their annual migration, journey three thousand miles to the south. But they always return to these same rocky islands to bear their young.

Ships bound from American ports to Bering Sea and the Arctic far to the north, with the opening of navigation in the late spring, set their courses for the passes in the Aleutian chain, beyond the end of the Alaskan mainland. Mariners with their instruments can steer unerring courses for these straits thousands of miles away. On such trips in the spring they find great herds of seals that have wintered in the south, off the coast of British Columbia or Washington, bound for these same straits that they may enter Bering Sea and roll on a few hundred miles to their breeding grounds. And no mariner with all his instruments ever steered a straighter course over these vast ocean waters than do these dumb animals, day or night, clear or cloudy, guided by some unexplained instinct within them.

These seals and their ancestors probably have been coming to the Pribilofs unbrokenly for hundreds of thousands of years. The Russian herds and the Japanese herds, likewise, have their own breeding grounds and their own feeding grounds on which they mix with no other seals. So long have they been thus keeping apart from each other that, although these seals were originally all alike, they have come to have certain fundamental differences, have come to be as different, for instance, as are the Japanese and the Americans.

But let us take a look at the greatest colony of them all, there on the Pribilof Islands so far away. It is a calm, gray day early in the month of May. The gloom of the long Arctic night has lifted. The snow is melting from the islands and innumerable flowers are hurrying out for their brief blooming in the short Arctic summer. The Pribilofs are serene in an unbroken solitude and one may circle each island and never see a member of

the herd which makes them famous, for there is not a
single seal to be seen hereabouts nor has there been for
six long months. These islands which are aswarm with
seals from June to October seldom see one of them from
November to May.

Now comes a solitary habitant of the deep, a great bull
seal weighing eight hundred pounds, sturdy and strong

AN OLD BULL SEAL THROWING OVER A YOUNGER ONE WHO TRIED TO
INTRUDE UPON HIS FAMILY

from many months of feeding upon the fish of the broad
Pacific, out there in the playgrounds of the herd. This
huge seal flounders up the rocky beach on his flippers, awkward and ungainly when he plays the rôle of a land animal,
although in the water he is a thing of unbounded grace.
He selects a spot on the beach covered with boulders and
jagged rocks. This is his idea of a place of comfort

and is to be his home and that of his numerous family during the busy summer months that are to follow. He has come early that he might take his choice of all the home sites on the beach, and this, the most rock-strewn, suits him best. Here he will locate his claim. He will become a squatter on this piece of Government land and will give battle to any intruder, man or beast, who trespasses upon it.

The next day there appears another bull seal and locates a similar claim. In a few days they begin to come in greater numbers and soon the whole beach is dotted with these huge patriarchs of the herd, braced upon their flippers and rearing their heads in defiance.

For a month these males sleep on their rights. Then the cows also begin to return to the summer home. It is strange that these seals should be called bulls and cows, for these are terms usually applied to grass-eating animals and these seals are flesh eaters. Since they are so-called it is even stranger that their young should be called "pups," like the young of d gs and wolves, but so they are known on the Pribilofs.

"Surely," the casual observer is likely to say as he sees the arriving females, "these are not the mates, but the children of these huge males!"

He is likely to reach this conclusion because the female seal averages less than ninety pounds in weight, while the males average five hundred. Almost nowhere else in nature is there so great a difference in the size of males and females.

With the return of the female the strategy of the old bull in coming early and staking out his claim begins to be understood. This beach upon which the males have located their homes is the breeding ground, and here the females have come to rear their young. The male that has staked out the most convenient claim and is able successfully to defend it will acquire the largest family, and big families are the dominant ambitions of these old seals. One such seal may control a plot of ground on which fifty, even one hundred females will gather. As long as he can control this plot these cows and their offspring belong to him. The old bulls may fight desperately

A SEAL FAMILY

for the possession of some newly arrived cow. After a cow has settled on the plot of a certain bull he guards her watchfully, and does not allow her for a moment to stray away for fear some other bull may get her. The old warriors get large families and the younger warriors get families that are smaller, in proportion to their skill and fighting ability. There are bulls that have but a single cow.

As it works out, this scheme of seal civilization leaves many of the males of the herd, in fact, the great majority of them, with no families at all. These are usually the younger or weaker males. They have no part in the family life of the colony. Yet they hang upon its fringe as the substitutes might be kept in reserve at a ball game, ready to be drawn upon if needed. These bachelor bulls are arranged back of the colonies, tier upon tier. In the front row are the ablest fighters of the reserve, ready to take the places of bulls in the rookeries that become disabled in the fighting. Back of this first line of reserve is a second and a third. Then, down on the beach, as a further addition to the colony, are yet other great numbers of its younger males. Finally, last of all to arrive, are the little seals, one and two years old, which play about the water front. Thus, by the end of summer, a full enrollment of the strength of the herd is to be found on the home grounds.

It is upon this scheme that the seal community of the Pribilof Islands is built. This plan of many homes might be spread out upon a given beach until all its space is occupied. Upon that beach there might be one hundred thousand individuals. On that beach during the summer twenty thousand young seals might be born. Around the corner is another beach with another such gathering and so on until all the seals are accommodated.

One mother of the twenty thousand in a given group might, after the time has arrived when the bulls allow them to do so, leave her individual pup there among the mass, flounder down through the maze of family plots to the water's edge, and go for a swim of one hundred miles into the ocean and for a day-long feast on fish. Then she would return to this complicated rookery, pick her way wigglingly on her stumpy flippers, progressing much as does the measuring worm, through the mass of individuals that compose the herd to which she belongs, showing no interest in the multitude of young ones strewn about until she finds that one little seal among them all which is her own and to which instinct guides her. This little seal she takes to herself and nurses fondly.

Through all this summer the old male stands guard over his flock. Not once through all those weeks does he leave it for a single moment. During all that time he does not take a morsel of food. Not once does he so much as go down to the water for a drink nor for a single dip into that water which is his native element, which is his natural home. He is a perfect example of watchfulness, faithfulness, and endurance. He yields to nothing, man nor beast. In the old days when the seal hunters raided the rookeries these old bulls stood their ground in defiance. After the raiders have gone, they have often been found dead, there at the threshold of the home plots, facing the enemy, defiant and unyielding. All the other members of their households have been skinned, but the hides of the warrior bulls, because of the many cuts they have received in battle, are worthless, so they have been left untouched.

This odd manner of the life of the fur seal herds causes them, like the herds and flocks of cattle and sheep, to lend themselves ideally to commercial purposes. Seals are polygamous: one male has many mates in his family, as is the case with most farm animals. Because of this fact the great mass of the males are not necessary in keeping up the numbers of the herd. It is as in the case of cattle raised for beef. Nine males out of ten may be slaughtered and the rate at which the herd will grow is not affected. The females will bear as many calves as they would if all the males were there. With seals it is as with cattle. Nine males out of ten may be killed, in this case for their skins, and the increase of the seal herds will not be checked.

So, the Government, in the operation of this herd on the Pribilof Islands, from the time it took charge of it, adopted the policy of killing for their skins only the surplus males. By the time it had been operating the herd for ten years it found that it might kill some thirty thousand male three-year-old seals each year without in any way interfering with the increase of the herd. This it proceeded to do, and thus it produces a great quantity of seal skins every year which it sells at a handsome profit, turning the money into the public treasury.

All the killing is done here on the Islands by agents of the Government. When the seal colony takes form in summer the young three-year-old males, from which most of the animals to be killed are selected, hang upon its fringe on what is known as the hauling grounds. The Government agents, with native helpers, have but to go to these hauling grounds and to drive away the number of males they want for killing as simply and easily as they could drive away a flock of geese. These awkward seals, traveling by land most clumsily, may be driven slowly for a mile or so to the points at which they are skinned, where oil is refined from their bodies, where disposition is made of the refuse.

As the summer season draws to a close the little seals, now about two months old, go down to the beach and shyly take their first dip in the ocean. They swim out a little way, then return, climb back into the sun. Presently they venture farther and then farther, and in a few weeks have found their proper home in the water. By this time the form of the colony ashore has pretty much broken up. The tired old males, now lean and gaunt from lack of food, creep into the grass and sleep without waking for a week. As autumn advances bulls and cows join the young ones and roll away to the south, traveling as far down the coast as Santa Barbara, California. They spread out over half the North Pacific, and feast and frolic endlessly until another spring has come. Then they drift back, with the lifting of the Arctic night, to their rocky, far northern home for those few weeks of each year during which, though awkward at it, they are none the less animals which live on land.

One is not likely to conceive of these water animals as being mammals that feed their young on milk as do those animals all about us with which we are most familiar. Yet this is true. The seal bears no relation to the world of fishes, but is very closely related to land mammals and particularly to the animals that prey, that live on flesh. The seal, properly classified, is most closely related to the flesh-eating land animals—the bears, the cats, and the wolves. Most of all is it like the northern bears which live much in the water and live on fish. Thus, with its cousins, the ordinary, widely distributed hair seals, which produce no fur, and the sea lions, does it fit into the animal world, there where it would be little expected, as a carnivorous mammal.

The fur seal comes to occupy also a peculiar position with relation to the Government. Ownership of this herd has put the Government into the sealskin business and the Government very rarely goes into any business at all. It has given the Government control of the sealskin business of the world, for ninety per cent of all skins that come to any market are produced by this herd.

When the Government had been operating the herd for a decade it had run its numbers up to six hundred thousand. On the basis of killing only the surplus males it had found that it was able to take thirty thousand skins a year. It had concluded that it might safely figure that it could kill each year a number that was five per cent. of the whole. On this basis the herd would increase in numbers five or six per cent each year. It figured that it would have a herd of a million by 1932, and in ten years more, if it saw fit to develop it, there might be a herd of two million. When it had a million it would be getting fifty thousand skins a year, and when it had two million it would get one hundred thousand skins a year. When it saw fit to stop the increase of the herd, to hold it at a given figure, it might, by killing some of the females, double the number of skins.

It has already been shown, however, that the action of the Government has saved to the world an interesting animal and a producer of wonderfully beautiful fur. There are fragments of herds elsewhere that might be similarly built up. Russia has a colony with possibilities. Uruguay has one near the mouth of the Rio de la Plata. Chile has the bare remnants of several herds that, with careful nursing, might grow. There are small groups left in the Antarctic. Only international action can save these seals, action that stops them from being killed

other than as they are killed on the Pribilof Islands.
The United States has demonstrated the method of saving sealskins to the generations that are to come.

QUESTIONS

1. What two kinds of seals are most important? Where do the fur seals live? Describe their strange attachment to one particular summer home.
2. What can you say of the number of fur seals a hundred years ago? twenty-five years ago? What is the situation today?
3. Tell the story of the seals of Pribilof Islands. What is pelagic sealing? Explain its wastefulness. Show how these animals were saved by a treaty.
4. Show how the United States built up the herd. How can it control the numbers of the herd?
5. Describe the coming of the seals to their breeding grounds. How does the bull seal establish his claim? Compare the size of the males and females. The defiant challenge of these old males is like that of the gladiators of the Roman arena. Describe the fighting that follows. Show how the idle bulls come about.
6. How long does the vigil of the bull seal last? Why could not other animals keep this vigil?
7. The fact that one male may have ten or fifty families causes many males to have no families at all. How may these males be used without weakening the herd? Show the commercial possibilities that lie back of this fact. How do the government agents select the seals to be killed?
8. Describe the breaking up of the seal colony in the autumn. How do the seals live through the winter?
9. Is the seal herbivorous or carnivorous? Is it a mammal or a fish? Is it a beast of prey? An examination of its anatomy leads to the conclusion that its ancestors lived on land and walked on four legs. What are its nearest relatives?
10. The United States Government has shown how a fur seal herd may be built up. Where are there other small herds? What possibilities lie in them? What is necessary that they may be developed?

CHAPTER XVII

THE HOUSE RAT

I N the fourteenth century in Europe the rat was responsible, through carrying the "black death," for the loss of the lives of twenty-five million people.

No longer ago than the year 1907 it caused the death in India through bubonic plague of two million human beings.

Each year in the United States a large amount of food is destroyed by rats. It has been estimated that as many men are required to produce that food as are enlisted in the army and navy for national defense.

Among those living creatures in the world larger than insects, the rat is the greatest enemy to man—is the unwelcome guest in his household which costs him most dearly. Insects, it should be remembered, are animals, and among them the fly undoubtedly causes more disease and death than does the rat. Insects, such as the boll weevil, do great damage to crops. It should be borne in mind also that even a germ is an animal, and that germs are responsible for many deaths, as, for instance, in tuberculosis. Insects and germs are man's worst animal enemies. Of the larger creatures, however, those which we are likely to think of as animals, the rat does man more harm than any other—causes more illness, costs more money. The rat is man's worst enemy among the vertebrate animals.

As one tries to lay out before him the map of the animal world, to find the place into which each creature fits, it is interesting to discover that each has its niche and is in that niche quite separate and apart from the multitude of all the others. The rat, for instance, is found to dwell among the rodents, and rodents are different because they are creatures that gnaw. A certain group of the animals in this world are furnished with teeth which fit them very effectively for the business of gnawing. They make their livings by gnawing. All these creatures are related. The rabbit, for instance, is a gnawer, as are the squirrel, the beaver, the porcupine, and, finally, the rat and the mouse. They are all mammals—are a division of the mammals, just as the flesh eaters and the grass eaters are divisions of that class. The chief orders of the mammal class may be expressed as follows:

MAMMALS		
CARNIVOROUS	HERBIVEROUS	RODENTS

There are more rodents than all other mammals combined. And for sheer destructiveness the king of the rodents is the old, brown rat.

It is interesting to know that when America was discovered there was not one of these brown rats in the western hemisphere. It is also interesting to know that Julius Cæsar with all his journeying, and Aristotle with all his wisdom never knew of the existence of the rat.

The rat as a creature of the western world is, therefore, a thing of quite recent times.

Before the coming of the brown house rat, now chief of man's rodent enemies, there was another race of rats, powerful in its time, that has almost disappeared in the presence of its more aggressive rival. This was the black rat which is believed to have come out of India about the time of the Crusades and which gradually spread west over Europe, reaching England during the thirteenth century. This black rat was the dominant species in Europe through the Middle Ages—was the creature to whom the Pied Piper played, and was the first member of this Old World race of rats to find its way into America. There were probably black rats on the *Mayflower* when it landed at Plymouth, and these probably came ashore and started colonies that increased more rapidly than did those of the Puritans.

The black rat held sway in rodent land of the western world for about five hundred years. It was this black rat which carried bubonic plague, and spread it over Europe during that century which preceded Columbus, where, in some countries, it killed as much as two thirds of the population.

In the rat history of the world the year 1727 stands out as the most important of all dates. It was in that year that the brown house rat that has since spread around the world, swam the River Volga, there in the middle of Russia, and started its sweep to the west. In a few years that sweep had covered all western Europe, and in 1775, just before America's Declaration of Independence, the brown rat was reported to be in existence

RATS HAVE A GREAT PROPENSITY FOR TRAVEL

on this side of the water, where it began to increase and multiply, to play its rôle in American life, to become a part of pioneer pilgrimages, to appear in California, for instance, for the first time with the vanguard of the Forty-niners.

The most remarkable illustration among the rodents of an instinct for migration is to be found among those tiny mouselike creatures, the lemmings of Scandinavia. They breed and become very numerous in interior Norway and Sweden. Then they set out in a march, east or west, millions of them, swarming over great areas of country. When they encounter rivers or lakes, they swim them. When they encounter mountains, they go over but never around them. Finally, when the survivors reach the ocean, they plunge in and carry on in their course. So they die.

The black rat pretty well spread itself all over the world because of its great fondness for playing the rôle of stowaway upon ships, for riding about, and for going ashore wherever it saw fit. The brown rat has the same instinct for travel and has repeated the experiences of the black rat. Thus it appeared in the same ships, on the same docks, in the same cities, the same households, with the earlier rodent pioneer. There the two met nose to nose and eye to eye. And in these contacts the heart of the black rat quailed and was stricken with terror. The brown rat was acknowledged the master.

And well it might be recognized as master, for did it not sometimes weigh a pound and a half, and reach eighteen inches in length, twice the size of the black rat? Had it not a will and a greed that stopped at nothing

save other larger animals that might terrorize it as it
terrorized its lesser cousin? Did it not show itself a
cannibal? Did it not, when quartered with this lesser
rat, slay and devour it? Did its womenfolk not give
birth to twelve families a year and did these families
not each contain an average of ten children? How could
any lesser race survive in competition with so dominant
a strain? As a matter of fact, the black rats were steadily
displaced wherever the brown rat came, and tended to
disappear entirely.

In 1904 one farmer in Mercer County, Illinois, killed
upon his premises 3,445 rats. The health authorities of
San Francisco in a campaign between 1904 and 1907 killed
and counted 800,000 rats, and New Orleans in 1914–1915
piled up 550,000 of them. On one rice plantation in
Georgia in a single year 17,000 rats were killed, while
a meat-packing establishment in Chicago does away
with 8,000 a year. Thus is it shown that the brown
rat is widespread and abundant.

Probably the most perfect example of the ability of
these rats to establish themselves in new lands and to
thrive under new conditions is given in the case of the
island of South Georgia in the Antarctic Ocean, an island
which is used as a basis for whaling expeditions. Whaling ships have brought rats to this island. The men have
killed many whales and left the bodies of some of them upon
the beaches. There these whale carcasses, kept from decay
by the cold weather, have become abundant food for
the brown rats from the far-away heart of Asia. In
the presence of this unlimited food supply they have increased in stupendous numbers until the island of South

Georgia has become the region of densest rat population in all the world.

The favorite abodes of the rat are the structures built by man. The house in which man dwells is a comfortable abiding place for the rat and is likely to contain an abundant food supply for it. The storehouses and particularly the barns he puts up offer opportunity for unlimited feasting. The docks of all his water fronts offer innumer-

THE RAT IS ONE OF THE CHIEF CAUSES OF FIRE

able hiding places; near-by warehouses furnish food, and the ships that come and go are for the rat regions of delight. The grains that man grows on his farms are nature's food for the rat, and in these it thrives and prospers. Even such articles as the eggs of his poultry, which come to market in cases, may be utilized by this cunning marauder, who will gnaw a hole in the side of a case and carry the eggs down flights of stairs to its burrow

without breaking them. These gnawing teeth of the
rat grow constantly and must be kept worn down. So
it will gnaw to keep its teeth in shape when there is no
other occasion for it. Because of this practice the rat
often gnaws the insulation off electric wires and sets the
house on fire. It even punctures lead pipe.

Man pays the board and damage bill for the rat.
It is estimated that in the United States today there
are more than one hundred million of these greedy
devourers of the food supplies that man provides for
himself and his beasts of burden. Each rat, it is figured,
costs two dollars a year. It is easy, then, to get an idea
of the tax upon any city that is due to the presence of
rats. Roughly it may be said that a city pays as much
to maintain its rats as it does to maintain its streets.

The great health menace of the rat is through the fact
that it carries bubonic plague and the germs of other
diseases. The rats have the plague the same as human
beings. Rats also have fleas. When a flea bites a rat
that has the plague it becomes infected with the
disease. Then the flea may leave this rat. It is certain to leave it if the rat dies. It gets on another rat
or on a human being and begins drilling a hole so that
it may get some warm blood for breakfast. It thus
plants bubonic plague germs in that hole. The new
rat or the human being gets the plague.

By passing the disease in this way from rat to rat,
the disease is kept alive indefinitely. When there was
plague in India in 1907, for instance, rats having it got
on ships and rode to many foreign ports, carrying the
disease with them. Every time one of them died of

plague, its fleas left it and got on other rats which they bit. Ships with infected rats came into San Francisco. Every effort was made to keep them from coming ashore, but despite this they got in. The rats of San Francisco got the plague. It spread to country rats near by. Fleas got from them to ground squirrels which carried it into their burrows in the open country. Imagine the difficulty of getting rid of it under the circumstances. There it remained for decades. Occasionally a ground squirrel is still caught that is infected with bubonic plague. Some time it may flare up again even in America.

So does it become easy to see that man should awake to the danger that lies in the presence of rats. Aside from this, he should realize that the presence of rats costs him money all the time—two dollars a year a rat. He should fight rats.

There are more rats in the world than there should be. Nature has got out of balance with relation to the rat. As usual, when Nature is out of balance, the interference of man is to blame. In Nature the rodents live largely on seed-bearing grasses and the carnivorous animals live on rodents. In accordance with the law of Nature, when the grasses become abundant, and food is plentiful for the rodents, they increase in numbers.

The abundant supply of seed-bearing grasses causes the development of an abundance of rodent, bird, and insect life that feeds upon it. They reduce the abundance of grasses by eating it up. Their food supply becomes short and they breed less rapidly. Thus is there a tendency to keep each in bounds, to maintain the balance of Nature.

There is yet another check of Nature one step farther up. The flesh-eating animals, as, for instance, the foxes, weasels, and others, depend largely on rodents for food. If foxes are very abundant, they will catch most of the rodents and their food supply will become short. In turn, they will starve or, at least, in the presence of a shortage of food, will breed few young. They will become scarce. This again will give the rodents a better chance and they will become more abundant. The foxes will then have plenty of food, and they, in turn, will become more plentiful and eat the too-numerous rodents.

In this way Nature, when left to herself, will keep the balance—will maintain the right proportion among her creatures.

Man, however, interferes. He has, for instance, developed grain crops, which are the chief dependence of rodents, far beyond the degree in which they grew in the natural state—has grown limitless rice fields, wheat fields, corn fields. This has given the rodents an unnatural abundance of food. On the other hand, he has slaughtered the carnivorous animals and birds, the hawks, owls, bobcats, foxes, weasels, bears, and snakes, and thus has removed many natural checks on the rodents. He has put up innumerable buildings that furnish places of retreat and safety for them, and protect them from their enemies. He has upset the balance of Nature. The rodents, particularly the rats and the mice, have become overnumerous—have become a menace. Man is put to the necessity of undoing his work. He must take a hand in keeping the rodents down. He must fight them.

The fight upon the rats must be organized, must be a community fight, a municipal fight, a county fight, a state fight, a national fight, a world fight. The first step in such a fight is to support those agencies which develop the scientific information, the technical detail as to how the fight should be carried on. Municipalities, states, and the Federal government through their health and agricultural agencies study such problems as that of the rat. These studies require financial support, the backing of the taxpayer. The expense to the individual taxpayer is almost nothing, but he must be willing to bear that expense or the work cannot be carried on. It therefore becomes the citizen's duty to support those agencies which study such problems as this.

When the problem has been studied and the advisable method of procedure determined, it then becomes the duty of the citizen to support and participate in the necessary measures to bring relief. Already many facts are known. It is known, for example, that the householder's tendency to rely upon the cat for the suppression of rats and mice is a case of misplaced confidence. Most cats do not even catch mice, and few of them will tackle a grizzly house rat.

An excellent example of this fact was developed by the health authorities in New Orleans when they insisted on fumigating the cabin of a ship captain who argued that it was not necessary because he kept a cat in it. The apartment was fumigated, however, and unfortunately Tabby was overlooked and killed. With her in the cabin, however, were found twenty-four dead rats with which she had been living in peace.

Another great mistake which tends to give the rat a better chance is in the slaughter of hawks and owls. Of all enemies of the rodents, the hawks and the owls are most effective. An occasional hawk and an occasional owl, however, preys upon the farmer's chickens. The farmer, therefore, concludes that all hawks and all owls are his enemies and proceeds to slaughter them all. He is little aware that one barn owl will kill hundreds of rats in a season and save great quantities of grain from waste. He is little aware that one hawk, not of the chicken-catching variety, will fly about his fields and work industriously toward keeping down the rodent pest.

The basis of man's campaign against the rat, however, lies in the construction of his buildings. All buildings should be rat-proofed. In the long run it is much cheaper in constructing a building to make it rat-proof. The farmer's barn, in which his grain is safe from the rats, may save him hundreds of dollars a year for a lifetime. The shopkeeper, the warehouseman, the mere house owner may save himself much money and aid materially in the suppression of the rat by the proper use of cement at his foundations and screening at his windows.

The detail of constructing rat-proof houses, of trapping rats, of poisoning them, of developing their natural enemies, of organizing for their suppression, cannot be gone into here. Full details and instructions can be secured by addressing a letter to the Biological Survey of the Department of Agriculture at Washington, inclosing ten cents and asking for a copy of the Farmer's Bulletin on rats.

QUESTIONS

1. Why is the rat the most harmful of all animals?
2. Into what three orders are mammals divided? To which of these do the rats belong? What is the one chief peculiarity of the rodents? How do their numbers compare with those of the other orders of mammals?
3. It is a peculiar fact that, a thousand years ago, there were no rats outside of the Orient. What breed of them came first? What great tragedy did it cause? Describe the westward sweep of the brown rat.
4. What were the traits of the brown rat that made it so successful in new countries? Show how abundant it has become in America.
5. Show how man has created a world peculiarly favorable to the rat. How many rats are there in the United States? What do they cost man?
6. What dread disease does the rat carry? Describe its method of spreading it. Has the rat ever brought it to the United States?
7. Show how Nature, left to herself, establishes a balance. Show how man and the rat have upset that balance.
8. What is the first step to take in fighting rats? What government agency is studying the rat problem?
9. Are cats a great help in fighting rats? What are the greatest rodent enemies? What mistake has man made about hawks and owls? Every member of this class could help in the fight against the rat by explaining, on all occasions, why the hawks and owls should not be killed.
10. Rat-proof buildings are the biggest help. Are the buildings in your neighborhood rat-proof? Where can you get reliable information on rat-proofing and other measures for the suppression of the rat?

Chapter XVIII

THE CRAWFISH

HERE enters a whole world of new animals—jointed animals with jointed legs.

Of course one knows the crawfish, its big cousin, the lobster, and its crustier relative, the crab. They are all jointed animals with jointed legs. They are the giants of their race. They have eight walking legs besides their nippers and their swimmerets. They are covered by thick crusts and are, therefore, called crustaceans.

But these specimens are somewhat unimportant as the jointed animals go. One is more likely to meet a spider in one's daily life, and it, too, has a jointed body and jointed legs. Or one may meet a scorpion which has a disposition like a spider, but shows its relationship to the crawfish very plainly in the way it is put together. It, too, has four pairs of jointed legs and two nippers.

Then there are centipedes with jointed bodies and jointed legs—great numbers of these legs. Centipede means literally "hundred foot," although that is not the exact number of these members they possess. There is the millipede, whose name means "thousand foot," when, as a matter of fact, it has only a few score. The bodies of these animals are divided into segments and

17   239

each segment has a leg on each side. The simplest of them have ten segments and twenty legs, while the most complicated have one hundred seventy-three segments and, therefore, three hundred forty-six legs. These, too, are of the jointed world.

Then, finally, there comes the biggest group of them all, the most numerous of all animal groups, the insects. An insect is an animal with a jointed body and jointed legs. It therefore belongs to the same broad classification as the crawfish. It is what the zoölogists call an arthropod, for "arthro" means "jointed" and "pod" means "foot."

So we go back to the main divisions of the animal kingdom, we disregard those with backbones, we consider the whole field of those without backbones and we find the jointed animals, the arthropods, tucked in among the rest in this way:

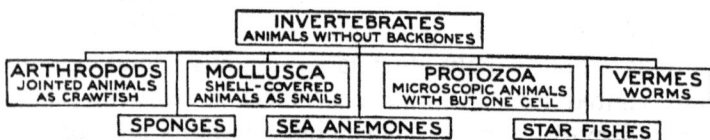

```
                    INVERTEBRATES
                 ANIMALS WITHOUT BACKBONES
  ┌──────────┐ ┌──────────────┐ ┌─────────────────────┐ ┌─────────┐
  │ARTHROPODS│ │   MOLLUSCA   │ │      PROTOZOA       │ │ VERMES  │
  │ JOINTED  │ │SHELL-COVERED │ │ MICROSCOPIC ANIMALS │ │  WORMS  │
  │ ANIMALS  │ │ANIMALS AS    │ │   WITH BUT ONE CELL │ │         │
  │AS CRAWFISH│ │   SNAILS    │ │                     │ │         │
  └──────────┘ └──────────────┘ └─────────────────────┘ └─────────┘
      ┌──────────┐    ┌──────────────┐      ┌─────────────┐
      │ SPONGES  │    │ SEA ANEMONES │      │ STAR FISHES │
      └──────────┘    └──────────────┘      └─────────────┘
```

Among these jointed animals the insect has one outstanding peculiarity that sets it apart from all other jointed animals. It has a jointed body with jointed legs just like the rest of them, but it always has a certain number of these legs, a number different from the rest. It always has six legs. Every insect in the world has six legs, no more, no less. Every animal in the world with six legs is an insect. No animal with any other number of legs is an insect. The spider, for instance, has eight legs and, therefore, is not an insect. The coral polyp has no legs at all and, consequently, cannot be an insect. But bees, flies, beetles, and moths have six legs and are insects.

All these jointed animals are related to each other just as all animals with backbones or all mollusks are related. The animals with backbones are most highly developed and, with man at their head, rule the world. The jointed animals are most numerous and are the most threatening enemy of the vertebrates. The insects, the scientists say, threaten the very existence of the higher animals. The present trend on the face of the earth is toward increasing strength for the insects and diminishing strength for the vertebrates. The insects may, some day, eat up all the food of the higher animals and thereby cause them to starve to death and disappear.

The crawfish is, with but few minor modifications, an enlarged insect. An ant, for instance, is built much as is a crawfish. The lobster is a salt-water close relative of the crawfish, and in general form much like its cousin of the inland streams. All the members of this world of jointed animals are built on the same general plan—the crawfish plan. A description of its make-up, therefore, in a general way fits all jointed animals.

The body of the crawfish has two major sections: the thorax, or trunk, which is the main part of the body, and the abdomen, which makes up the second division. In these jointed animals the head and trunk occupy the same division, there being no neck.

The jointed nature of the members of this phylum, or

branch, of the animal kingdom can be most readily seen by examining the abdominal or back part of a crawfish or lobster. It will be found that this division of the animal consists of a series of six rings, each jointed into its fellow and capable of considerable movement at the joints. An examination of the trunk would show also that, while the top of it is covered with a hard and unbroken shell, the bottom is divided into segments and is capable of movement. The body is thus jointed throughout.

THE CRAWFISH LIKES THE RIVER BANK WHERE ROCKS AND MUD ARE MASSED TOGETHER

The crawfish has four pairs of jointed legs, all of which are attached to its trunk. These legs are like so many links of sausage hitched together, except that the links may be of different lengths and the outer covering is hard. It uses them very little for any other purpose than walking about there on the bottom of the stream in which it lives. The two pairs in front have nippers on them with which it may anchor itself and which it may use at meal time to serve the purpose of forks in carrying food to its mouth.

In front of the four pairs of walking legs of the craw-fish are its claws or hands, which are nothing other than an excellent and handy pair of forceps which it uses for catching game and supplying itself with flesh for food. The nippers of these forceps are very much like those that one carries in the tool box of one's automobile, and they work in the same way. They are good for grabbing and holding on. There is great strength in them. The toothsome morsels of white meat that one finds in the nippers of a lobster or a crawfish are nothing other than the muscle which operates these forceps and gives them their surprising power.

On each ring of the abdomen of the crawfish there is a little leg known as a swimmeret, while at the end of the tail are four broad paddles or flippers which are the chief reliance of the crawfish when it wants to move quickly under water. It is these flippers, for instance, that en-able the crawfish to perform that act for which it has be-come famous, the act of moving backward. With its tail curved under it these flippers at its end are normally pointing forward; by moving them vigorously when in this position its body is forced backward. The animal "crawfishes."

The crawfish has two prominent eyes protruding on stalks at the front of its head. These eyes, like all those of the jointed animals, are built on a plan peculiar to them. There is no eyelid, but the eyes stand out there like skylights, open all the time. There are no eyeballs or pupils. There are merely lenses for letting in the light like those used in cameras. There are many lenses in a single eye. They are packed in closely like the cells in a honeycomb. They are so small that there may be thousands of them in the eye of a single insect like the house fly. These lenses point in different directions, so that the crawfish or fly may see all around without turning its head or moving its eye.

The crawfish has two antennæ or feelers, half as long as its own body, sticking out from the front of its head. These are very sensitive to whatever is going on. It has two lesser feelers which hear and smell. At the base of these feelers is a hole opening into a cavity and protected by tiny bristles. This is the ear of the crawfish. There are always a few grains of fine sand in it. The crawfish must select these grains of sand and put them there

FRONT VIEW OF THE CRAWFISH SHOWING HIS LARGE HEAD, FEELERS, AND NIPPERS

after it sheds its skin or the ear cannot do its work. In some of the insects all these senses are combined in a single pair of antennæ, very remarkable instruments, which feel, smell, and hear, all through sensitive spots along their joints.

Aside from their joints, the second outstanding peculiarity of the members of this branch of the animal kingdom is the fact that they wear their skeletons on the outside. They are all covered with a crust or shell. This crust lends to their bodies the stiffness which is furnished by the bones of the higher animals. It is, in fact, their bones. Among the higher animals the bones are put inside, and the muscles, working as pulleys, are attached to them on the outside. In the case of the arthropods this shell is the bone, but it is placed on the outside, and the muscles working as ropes and pulleys to operate the parts of the body are attached on the inside. In this way the jointed animals, operating on a different principle from the backboned animals, may lay claim to an advantage, since the hard portions of their bodies on the outside protect the softer portions which are inside. From the crawfish's standpoint, man would probably seem very foolish to put his tender flesh there on the outside where it can be scratched by any briar or punctured by any hungry mosquito.

This outside skeleton of the jointed animals is usually made of chitin, a material very like our fingernails, but in the case of the bigger members of the race—the crabs, the lobsters, and the crawfishes—carbonate of lime, such as the mollusks use in making shells, is added.

There is one great disadvantage which the crawfish and all the members of its race experience in holding to this scheme of wearing the skeletons on the outside. They find that this hornlike covering will not stretch to allow them to grow. They begin as little fellows an inch long, put on this coat of mail, incasing their bodies in an ironclad way, and grow inside all of it until it is tightly filled, until they can grow no longer. This happens in the case of the young of crawfishes, spiders, grasshoppers, cockroaches, and all the rest of them. In this emergency there is but one escape—the growing animal must discard its jacket, which is too tight, and get another which will give it elbow room. It must do this not once, but at many stages of its development. In the summer time the old clothes of insects are to be found on every bush.

Under these circumstances here is what the crawfish does. When its case gets too tight it swells out its muscles inside it until the case bursts. It bursts, of course, at its weakest point which, with the crawfish, is underneath at the junction of the trunk and the abdomen. This is different from the insects, for they burst their old jackets down the back.

The crawfish, however, finds this business of shedding its skin to be a very serious undertaking. The hard jacket covers every portion of its body to its very tips; all those legs, swimmerets, nippers, and feelers are covered with it. The muscle inside of them is but flabby and formless without these stiffening cases. The crawfish faces the necessity of getting all its formless parts outside this coat of mail and then making itself a new suit.

Getting outside of its shell is the hardest piece of work that the crawfish ever does. It takes this work very seriously. One may watch it in the process. It may be seen to take one member after another and shake that member very vigorously. What it is trying to do, undoubtedly, is to shake the muscle loose from the shell on its inside. Having done this, it contracts the muscle and pulls it out of its sheath.

The crawfish seems to experience a good deal of difficulty in accomplishing this feat. It sometimes takes hours of hard labor and the animal will lie upon its side and even turn upon its back in its twistings and exertions. Its conduct is not unlike that which may be seen on the stage, when a performer is strapped into a strait-jacket and gives a demonstration of his ability, by twisting, squirming, straining, to wriggle out of it.

Finally, the crawfish emerges through the hole in the bottom of its case. Its shell is left there on the bottom of the stream in all its completeness, to all appearances a crawfish, though, in reality, but an empty shell. The actual crawfish which came out of it is a strange, almost formless creature, capable of little motion and filled with many terrors. The crawfish protected by its shell is a bold swashbuckler of the river bottom, but the crawfish without its shell is timid, shrinking, and terrified, seemingly aware of the fact that it is now an easy victim to any enemy which may attack it.

As embarrassed as a boy whose clothes have been stolen while he is swimming, it hides away there in the bottom of the stream for two or three days. Its chief business is to absorb water and swell itself out to a size much greater than before. This greater size is not brawn and tissue, but is due merely to the presence of water. What the crawfish is trying to do is to make itself big against the formation of another shell, for it wants this shell to have room in it in which it may grow. With its body thus swollen out the new shell which it secretes begins to take form, to harden. In two or three days the soft animal is covered with a radiant coat of new armor and again resumes its swashbuckling. It has a great, roomy house in which it may grow for another six months, replacing the water with which it made itself big by good, hard tissue and muscle.

An interesting accident may happen in the course of this process of changing shells. The crawfish may lose one of its legs or one of its claws. It may pull a leg off trying to get it out of the shell. Shedding its shell is serious business and it will pull a leg in two if necessary, just as it will tear off one of its nippers if it becomes necessary in escaping capture. Often one sees a crawfish with a leg or a nipper gone. The crawfish here again has an advantage over man because, when it loses a leg, it can grow another one. Watch this injured one for a few months and you will find that again it is fully supplied with its members.

The crawfish begins life in its own peculiar way, there in an egg which its mother lays and fastens to the swim-merets beneath her abdomen. There may be two hundred of its kind, for so many fall victim to their enemies that these creatures of nature start many young ones on the perilous journey. After they have hatched the little ones, a quarter of an inch long, stay close to their mother for a while, then venture timidly forth, only to hurry back into the crook of her body when danger appears. Finally, they are out for themselves, active little fellows on the bottom of the stream.

By repeated shedding of their shells the crawfishes grow at the rate of something like an inch a year until they become mature at the age of seven or eight years, and are seven or eight inches long. And, if they escape the pitfalls, they may live to the ripe old age of twenty.

With its coat of armor in place the crawfish is the stealthy forager of the banks and bottom of the streams. It likes a river bank where rocks and mud are massed together. This enables it to take refuge in the safety of an overhanging rock with only its nippers protruding. If there are not rocks available or roots of trees the crawfish will dig itself a cave in the river bank. There it will sit with its feelers sticking out into the water and with its nippers folded confidently beneath its chin. These feelers are the lookout stations which telegraph back the situation, which inform headquarters of the approach of any possible victim.

The crawfish is fond of eating such tender inhabitants of the water as the grubs of May flies or dragon flies. It can take the shell of a snail in its pinchers and crush it. It is said even to be able to crush a clam shell so that it can devour its contents. Little fish or frogs are to it a delicacy. Any dead thing of the stream is food to this crawfish, which is as much a scavenger as it is a huntsman. Yet, as a huntsman, the crawfish is not without prowess, and has been known even to capture a water rat, which is its arch enemy. The water rat, diving down in search of crawfish, may pass too close to the door of a great fellow that is lying in wait. The nipper of the big crawfish may flash out and catch the rat, and hold it, as in a vise, there under the water until it is drowned. Thus does the crawfish lay in a supply of fresh meat furnished by none other than its worst enemy.

Country boys know the habits of crawfish, crawdads.

they often call them, and know how to catch them. They have learned that they are good to eat, and they know how to make them ready for eating. The negroes of the lower Mississippi like them next to opossum. Both negroes and boys go crawfishing often, carrying a short pole with a string at the end, a piece of liver or frog meat tied to it, a net made of a barrel stave and a flour sack. They dangle the piece of meat before the crawfish burrow until it grabs it with its nippers. Before the crawfish can make up its mind that it is foolish to hang on and can release what it thought was going to be a dinner, the boy has pulled it within reach of the home-made net and it is captured. The process is repeated and soon there is a bucketful of prizes. In the meantime a weed fire has been burning under a pot of water and all the victims are plunged in and boiled, and thus made ready for a peace offering to mother when the boy gets home late to dinner.

In France the crawfish, like frog legs, is more highly prized than over here. There are crawfish farms where they are raised for the market. The English, although crawfish are abundant over there, never think of eating them. They are scattered pretty well over the world. There are many varieties of varying sizes, but, as a matter of fact, with few material structural differences.

Odd fellows, these jointed animals, with their many tribes. There are among them brilliant-hued butterflies that flit about summer meadows, poisonous tarantulas from the tropics, bees busy with their honey making, tiny aphids that exist by the millions on the rose bushes in your garden, shrimps that are so numerous that they

cloud parts of the ocean, quaint water boatmen in the
ponds, devil's darning needles that mimic the plants on
which they live, tight-clinging barnacles on the bottoms
of boats. They all have colorless blood, no lungs, hearts,
or brains. Strange neighbors these.

QUESTIONS

1. What does the word "arthropod" mean when you take it apart?
   Arthropods make up a large branch of the animal kingdom.
   What are some of the animals that you know that belong to it?
   Is a grasshopper an arthropod? is a lobster? a mosquito?
2. The peculiar thing about the insect group is that its members have
   six legs. Is a crawfish an insect? is a spider? is a butterfly?
3. Has the crawfish a backbone? Is it, then, a fish?
4. Examine a crawfish or lobster and observe the manner in which
   it is built up of plates joined together. How many legs has it?
   Describe its nippers. Its flippers. How does the crawfish move
   backwards?
5. The eyes of the crawfish, of all the jointed animals, are built on a
   peculiar plan. Explain how they work. Why does the crawfish
   put grains of sand in its own ears? What does the crawfish do
   with its feelers or antennæ?
6. Where do these strange members of the animal kingdom wear their
   skeletons? We wear ours on the inside and put the tender flesh
   on the outside. In what way is the crawfish's scheme better?
7. How do members of this branch of animals change their skins?
   Can you find some of the old clothes of insects? Tell of the difficulties the crawfish encounters in getting a new shell.
8. How many little crawfish are born at a time? What does this number
   indicate as to the life chances of the crawfish? How much a
   year does a crawfish grow? How long does it take it to get its
   growth?
9. Describe the method of the crawfish when it goes hunting. What
   makes up its food supply?
10. How do country boys catch crawfish? Are they good to eat? Where
    are crawfish farms to be found?

CHAPTER XIX

THE ELEPHANT

HEN, on the Fourth of March, 1921, Woodrow Wilson and Warren G. Harding rode together from the White House to the Capitol, where one was to give up the office of President, and the other was to take over its duties, there was much surmise as to what they said one to the other.

Long afterwards I learned what they talked about. It was not of the great office nor of the welfare of the one hundred million people under its sway, but of the animals each had known and loved.

Some incident caused Mr. Wilson to ask Mr. Harding which of the dumb creatures he liked best. The latter said that he gave first place to the elephant, that the elephant had always seemed to him the most intelligent, the most human, the most kindly of animals. He thought of a story of an elephant that had been told him by his sister who had spent years as a missionary in India.

It happened at a little village, lost among the jungles of the far interior. An old, faithful, thoroughly domesticated elephant had lived and labored long in this village and had come to be much loved by the inhabitants. By day he toiled diligently with his keeper piling logs of teak, that hard wood of India, and by night he stood guard against any marauders of the jungle that might seek to disturb the settlement.

Then the day of tragedy came. A great stack of the teak logs collapsed and tumbled down upon the elephant. He was bruised and crushed to such an extent that it soon became plain that he must die. The elephant himself seemed to know that the end was near. Gropingly he reached out his trunk to the native keeper who had been a much loved companion through the years. This strange arm of the wounded beast was wrapped around his man companion, who was nestled in a close embrace and there remained for five hours until the end came.

Well may the elephant be considered as taking first rank among interesting animals, for it is a curious, contradictory sort of creature, quite out of place in a modern world in which there exists nothing else at all related to it.

The elephant may well lay claim to being, next to man, the most intelligent of animals. There is no other animal, for instance, which can be taught as can the elephant to perform a task so constructive as building a tower of railroad ties, one layer running in one direction, and the next layer at right angles to it. There is no other animal that will work with one of its fellows with an intelligence which compares to that shown by two elephants, one at each end of a big log. There is no other animal which can perform a task so knowing as that of fastening a chain about an object with which to drag it away.

On the other hand, an excellent case can be made out on the theory that the elephant is the stupidest of all wild creatures—is without even ordinary instincts for self-preservation. Oddly enough, it is the elephant in the wild state in which it has lived through the ages that seems unusually stupid, and it is the tamed elephant, new to its strange surroundings, that shows most intelligence.

The elephant at home, for instance, roams the woods, congregates in deep forest shades, and shuns the sun. It is a sociable creature in which the family group is strong. Elephants live in herds, but the herd is usually the family. At the head of herds of horses, cattle, and most other gregarious or sociable animals there is always the fighting male which is the master. At the head of the elephant herd, however, there is always to be found some wise old mother who leads it.

The stupidity of elephants in the native state is shown by the ease with which they may be captured. Elephant hunters, for example, may find a herd of twenty or fifty or one hundred animals in some forest solitude. The hunting party, which may consist of two or three hundred men, may surround this herd in the forest and crowd in upon it until it has its members huddled together. Then a weak bamboo fence may be built, which a mere donkey would hold in contempt, but which this herd of wild elephants will respect. With such a fence, reinforced by an occasional watcher who may build a fire and wave a torch to confuse and terrify these great but timid beasts, they may be held in this circle indefinitely. They may be held, for example, until a stockade is built for purposes of more securely inclosing them. This stockade is not unlike the cattle corrals of the West, but is stronger. Wide wings are thrown out from it which guide the animals into its gate. By dint of great noise, by beating the brush, by flaring torches, the natives drive the elephants into the stockade, after which the gate falls and they are captives.

In the pursuit of the herd a queer thing seems to happen. As it grazed, there were a number of baby elephants to be seen. As soon as it took fright, however, they disappeared. They were not to be seen for the rest of the drive. They are not to be seen even after the herd is in the stockade. A careful examination will show, however, that they are still with the herd. They are, in fact, underneath the stomachs of their mothers. This is a usual happening. The baby elephants stay underneath their mothers, even while the herd is in flight. They can run as fast as a horse, and never be stepped on, there between their mothers' front legs.

Now come the trainers. They ride into this stockade on other elephants that have become thoroughly domesticated. These tame elephants are remarkably subject to the wills of their masters, and enter with enthusiasm into the task in hand. They appear to take great pleasure in helping to discipline the wild elephants. Under their protection the natives are able to fasten chains about the legs of the prisoners, are able to drag them forth, and fasten them to trees where they are left for days to fight it out with their bonds and become sub-

18

missive. They acquire the idea of submitting to their captors more readily than does any other animal. Later, yoked to tame elephants, they learn to perform the tasks set by man more readily than does any other wild creature. No other animal has the elephant's desire to learn, to do just what the master wills.

An elephant will pile stone in a wall. It will put a stone in place, walk away for a look at it, return, and adjust it. It will walk behind an army artillery piece drawn by oxen, and will steady it over a rough road to keep it from overturning. When it is stuck in the mud the elephant will lift one wheel and then the other to loosen it, will put its head against it to aid the oxen in getting it out, will give the signal for all to pull together. If the oxen fail to pull, the elephant may fly into an ungovernable rage and charge them threateningly.

No creature is more emotional—more manlike in its conduct. The elephant may grieve and refuse food almost to starvation when it is first captured or when its master dies. It may develop a fondness for the family to which it is attached that equals that of any dog. The natives give it tasks so delicate as that of tending the baby placed on the ground before it, a task in which the elephant takes obvious pride and to which it sticks, hour after hour, probably fanning its charge with the leafy branch of a tree.

But, occasionally, an elephant goes "must." This means that it becomes quite mad, vicious, destructive. Although it is usually the most timid of creatures, sometimes it is seized with the desire to kill. By sheer force of size and weight it is the monarch of the forest, but it

THE INDIAN ELEPHANT IS THE EASIEST OF ALL ANIMALS TO TAME AND TRAIN.

also has skilled methods of using its strength. It can strike and gore with its tusks. Its trunk can be used for catching and holding its enemies. Its plan of battle is to throw them upon the ground, to trample them, to kneel upon them and crush them. A mad elephant is a dangerous killer. Sometimes beasts that have been tamed for decades, that have lived in circuses and in zoölogical gardens, go "must" and there is safety only in killing them.

In the wild state there is an occasional elephant that becomes an outlaw, a "rogue elephant," as it is called in India. The rogue is a solitary male elephant. He leaves the herd and stays by himself. He may be a tame elephant that in rage has become an outlaw—has revolted against some such thing as the goad of his driver too severely used. His special delight is the destruction of cultivated crops, of dwellings, of human life. These rogue elephants select certain stretches of country as their hunting grounds and in that area no village is safe. They seem to have turned upon man, and to have determined to avenge themselves for the injuries man has heaped on them and their kind. They run riot until killed.

Hunting down these rogue elephants is choice though dangerous sport and is now almost the only elephant shooting left in India, as the government carefully protects the herds from destruction, supervises trapping them, and issues only occasional special licenses allowing one to be shot.

There was a time when elephants were extensively used in war, when great numbers of them, bearing down on the enemy, terrorized the warriors and won battles. Hannibal used elephants when he invaded Italy. But when the enemy comes to know the nature of the elephant, he may turn it into a destroyer of its owner. The elephant is at heart timid, given to fright, to terror, to stampede. If it is stampeded and driven frantically back through the ranks of its masters, it is likely to demoralize the ranks from which it came. The British, in India, no longer take elephants into battle.

In Burma the industrial elephant is seen at its best. There a single lumbering company cutting teak logs from the vast forests may have as many as one thousand elephants, each worth nearly two thousand dollars. It is no uncommon sight to see elephants arriving in Burma to replenish these herds, a hundred of them aboard a queer elephant ship built for the purpose of transporting them. Even more imposing are the processions of elephants kept by the native rulers as symbols of wealth for the purpose of impressing their subjects. But the idol of the western world is the stately elephant that marches solemnly about American streets when the circus comes to town and gives evidence under the big top of an ability to learn tricks that is beyond any other animal.

The elephant is the only animal among the mammals which the scientists give an "order" all by itself. The mammals, the milk drinkers with man at their head, are the dominant animal group of the world. In biology they form a "class." This mammal class is divided into "orders," the most important of which may be shown thus:

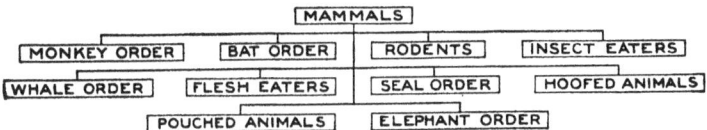

MAMMALS			
MONKEY ORDER	BAT ORDER	RODENTS	INSECT EATERS
WHALE ORDER	FLESH EATERS	SEAL ORDER	HOOFED ANIMALS
	POUCHED ANIMALS	ELEPHANT ORDER	

Animal orders are usually divided into families of which there are many. In the order of flesh eaters, for instance, there are the cat family, the dog family, and the bear family. Of the animals with trunks, however, there is but one family, the elephants. Families are usually divided into many genera, as, for instance, the wolf genus and the fox genus of the dog family. In the elephant family, however, there is but one genus. Genera are usually divided into species, and among the elephants there are two species, the Indian elephants and the African elephants. This is the only division in the whole line. These two cousins have no other relatives in the world.

There was a time on the earth not so long ago when elephants were scattered from pole to pole and lived on all the continents except Australia. They were quite abundant in America. A little earlier there was a creature closely related to the elephant, about twice as big and covered with long hair, called the mastodon. It is now extinct. Remains of it have been found imbedded in the ice of the North so well preserved that the meat was eaten by dogs and foxes.

The elephants and the mastodons are left-over specimens of the kind of creatures that ran about the place when the world was much younger. Most of the other big creatures have passed, and on land only the elephant remains. It is tending rapidly to disappear. Where a comparatively short time ago it lived all over the world, it is now to be found only in the wooded areas of southeastern Asia and in Africa south of the Sahara Desert. Even in those sections its numbers have been rapidly de-

THE AFRICAN ELEPHANT

creasing since man developed the repeating rifle and went forth to slay big game.

The Indian and the African elephants have lived in their different parts of the world for many thousands of years without mixing. As a result they have developed certain differences. In the first place their dispositions are different. The Indian elephant is much easier to tame; while the African beasts are wilder and fiercer. Nearly all the elephants that one sees in zoos and circuses are Indian elephants. The African elephants have larger ears that stand out more from the head. They are likely to be twelve feet tall against ten feet for the Indian elephant. Their tusks are twice as big as those of the Indian elephant. In India a tusk that is nine feet long and weighs one hundred pounds is near the record. In Africa they are found ten feet long, weighing two hundred pounds. These tusks are of ivory; and hunting elephants for them used to be an important business in Africa. It is a business that is disappearing through the rapid killing off of the elephants.

Since the Indian elephant is the easiest of all animals to tame and train, it would seem that these creatures would become domesticated and raised by man as are cattle and horses. This, however, has not been done for two reasons. In the first place elephants, although they seem to become quite happy in captivity, do not, except in rare cases, give birth to their young except in the jungle. In captivity they usually remain without offspring. So it would seem that, deep down in their natures, there is a great protest against this servitude to man with which they seem to have become reconciled. The second reason why they are not reared in captivity is that man is not interested in raising them and so has made little effort to overcome their tendency to continue without offspring. It takes an elephant thirty years to become grown. That time is so great that if a grown man started to bring up a baby elephant he would be likely to be dead before it came to maturity. It is easier to go into the forest and capture an elephant already developed than it would be to wait thirty years for one to grow and in the meantime try to provide the large quantities of food it would require.

The elephant is of an order, a family, and a genus of animals that has a proboscis, or trunk. It is the only order, family, or genus in the world that has this sort of member. This trunk is a thing to itself, unmatched in all creation.

And it is a very handy sort of thing to carry around. It is, in reality, a lengthening of the upper lip. The horse has an upper lip that is quite useful, for with it it can grasp grass blades and pick up grain in its trough. The trunk of the elephant is, of course, much more highly developed than this. It is a much handier arm than man's arm because it can be as readily bent as a hose at any point and in any direction. It can be used for grasping, for feeling, for striking, for breathing, for smelling, for trumpeting. The elephant can pick up a pin or untie a knot with it. He can wade in water a yard over his head, hold up his trunk, and breathe through it. It is quite sensitive and, in combat, the Indian elephant rolls it up neatly back of his tusks out of danger of injury. There are those who hold that the peculiar mechanical intelligence of the elephant is due to the fact that he possesses so handy a member as a trunk. The very possession of it would, through the ages, lead him to think of ways of using it, and this thinking would develop his mind.

QUESTIONS

1. How does the elephant compare in intelligence with the other animals? Point out some of the remarkable things it has been trained to do. Do you think it is more intelligent than a dog?
2. On the contrary, it does some very stupid things in the wild state. Point out some of them. Describe the life of the elephant in its native haunts.
3. How are herds of wild elephants caught? How are the individual elephants tamed and trained? How do the tame elephants help?
4. Tell how the elephant shows emotion. What is meant by an elephant going "must"? What is its method of fighting? Do you know any stories of "rogue elephants" in India?
5. How is the great strength of the elephant employed by man? in industry? in war? For show purposes?
6. Biologically to what "class" do the elephants belong? What are some of the important "orders" into which the mammals are divided?
7. The elephant is the only animal in the world that has a trunk. Are there many families of elephants? Many genera? Many species? What are the only two kinds of elephants in the world?
8. Where are wild elephants to be found? Were there ever elephants in America? Are the elephants a decreasing or an increasing, an old or a young race? Give reasons for your answer.
9. The same breed of animals, long separated from each other, will develop differences. Show how the Asiatic and African elephants have become unlike. Compare them as to size; as to disposition.
10. Have elephants been domesticated? What is the difference between being tamed and domesticated? Why have they failed to become domesticated?
11. Describe the elephant's distinguishing member, the trunk. Compare its usefulness with that of man's hand. How might the possession of such a member affect an animal's intelligence?
12. What effect has that master instrument of death, the repeating rifle, had, in the past generation, on the number of elephants in the world?

CHAPTER XX

THE GARDEN SLUG

KNOWLEDGE is like odd bits of iron or lumber, remnants of cloth, spare plumbing joints or electric-light sockets, bolts, nuts, screws, nails, pieces of string. Store it away and sooner or later a chance will come to make good use of it.

Knowledge of the unusual animals of the world would not appear to have much practical value. Those animals themselves would not seem to be of any use. It looks as if they might as well not exist. Studying their peculiarities, their habits, their life cycles appears to be a waste of time.

But he who holds that knowledge of any sort, of any living creature or physical thing is ever useless should remember that precious radium comes from useless shale, that iron and steel are taken out of ordinary red dirt. He should know the story of the soft and spineless garden slug, hidden beneath the garbage pail, and how it fought in the Great War to as good a purpose as marksmen firing long range guns or aviators dropping bombs.

Most people have turned over rotting boards lying in damp places and found garden slugs sleeping the day away beneath them. As might be expected they are "sluggish" animals, damp, slimy, and repulsive; stupid, slow, and helpless. They are two or three inches long, occasionally seven inches long, tender and watery, sleek and spotted. They make a soft and digestible item of food for the patient toad, occupying the place in its diet which watermelon supplies for man.

But the slugs are not so inactive by night. When it is dark they leave their hiding places for purposes of foraging, food hunting. They creep about the garden, eat fresh and tender plants and, therefore, sometimes injure the lettuce crops, the young pansies, or cucumber plants. The warmth and dampness of hothouses just suits them and there they sometimes become a nuisance, but not a very serious one. They are fond of the contents of the garbage pail, particularly of potatoes, custards, milk, cooked meats. They like man's food. They have rasping tongues with which they scrape their food off the objects they attack.

GARDEN SLUG CLIMBING A WALL, WHILE A SMALLER ONE DINES ON A STRAWBERRY

The garden slug is a very tender-bodied animal, without backbone, without shell. It is because of its soft and sensitive body that, wherever it travels, it builds itself a road to ease the going. It carries the road-building material for this purpose within its body. Man may use cobblestones, concrete, asphalt, whatever he likes, but the slug will lay its road as it goes, building it of slime that it manufactures for the purpose in its own movable plant. For every journey it lays a new roadbed.

Whoever examines, on a summer morning, the old stone wall at the foot of the garden, may trace the wandering of the slugs of the night before by the slime roads they have built, roads that are now silver ribbons in the sunlight. Taking a lesson from the slug some philosopher has written a motto: "Go slowly, go surely, and you will leave a brilliant trail."

The slime smooths the way for the slug. Wherever its system of road building cannot be used the slug cannot go. Put it on a surface which it cannot coat over with slime and it is as helpless as is an automobile in a sea of black mud. Put ashes, or flour, or even dust, in its path, for example, and it is immediately in serious trouble. When it begins to spread its roadbed of slime on these materials it fails as an engineer. The foundation for its roadbed is likely to give way and it may be plunged broadside into the ashes. This sticks to its slimy body. It flounders about and the situation grows steadily worse.

In this emergency it resorts to its one means of defense. It secretes more slime, surprising quantities of it. This does not solve its problem, but it goes on and on secreting slime for this is the only thing it knows how to do. It keeps secreting until it quite exhausts itself, until its soft body has wasted away in the process. Finally all its moisture is used up. Its skin becomes dry. It cannot live with a dry skin. It dies.

This fact that the slug cannot get along in the presence of ashes or dust is a good thing to know. It is a bit of unusual knowledge that may come in handy. Wherever garden slugs become a nuisance it is only necessary to surround their haunts with ashes to get rid of them.

Twenty years before the war a young scientist in the government service began what seemed like a most unprofitable study, an inquiry into the life and habits of the garden slug. Among other things he made flashlight photographs of them while they were on their nightly rambles, thus recording many of their actions during the hours of darkness. He kept them in a box in the basement of his house, where he had a parrot and a pet squirrel that were great playmates. These latter finally broke the plate of glass over the slug box. The prisoners escaped, and hid beneath the wooden floor where they thrived and multiplied.

It was after this that the scientist discovered the fondness of the slugs for boiled potatoes. At an earlier date slugs had been observed at night on a lawn. Some food had been left on a plate where a dog had been fed. A slug was seen creeping slowly toward that plate. The plate was moved off at right angles from the direction the slug was going and it immediately changed its course. By moving the plate the slug was led all about the lawn. Yet, there in the grass, the slug certainly could not see the plate. How, observers wondered, did it know the direction of the food?

Now this young scientist observed that wherever a

bit of boiled potato was left these slugs would find it and devour it. As is the way of scientists he began experiments to determine the slugs' method of finding boiled potato. Wherever he placed it the slugs would go directly to it. The experiment reached its height when he left boiled potato on the dining-room table. To get to this potato the slugs must come out of their basement hiding place, cross one room and a hallway, cross the dining room and crawl up the table leg. This they did unerringly.

As a result of many experiments this scientist decided that it could be nothing other than a sense of smell that guided these slow-moving animals. To be sure, cold potatoes have little or no odor as far as its detection by the nose of a human being is concerned. The observer knew, however, that many of the lower animals, particularly the insects, have more highly developed senses of smell than has man. Experiments have been tried, for instance, with the luna moth. Holes have been punched in its wings for purposes of identification, it has been taken three miles away from its mate and released. It comes straight back. Cover its antenna, through which it does its smelling, with flour paste, and it is lost.

Many of the lesser animals have this sense of smell developed to a degree that is almost unbelievable. Most of them, at the same time, have very poor eyes. It is their "noses" that guide them to their food and their mates.

Finally, in the midst of the war, a call came to the scientists in the government service that gave an opportunity to make use of certain queer information. The War Department said that a great emergency existed. The need of the moment was a method of telling when gas was present on the battle front. The Germans were sending over poisonous gases. They were killing and injuring great numbers of soldiers. They were sending over shells that were loaded with camouflage gas, gas that would do no harm whatever. But troops in the trenches could not tell whether this was poison gas or fake gas. They had to take all the precautions in the presence of fake gas that they did in the presence of poisonous gas. They were worried and harassed. They had to wear their gas masks alike for real and camouflage gas. They were uncomfortable in these masks—could not eat, drink, or smoke. This was hard on the morale. A body of troops that had been harassed for three days was in no condition to put up a good fight if the enemy came over the top.

. How, the War Department wanted to know, could troops at the front know positively when there was poison gas present and when there was not? How could they know when they should take the precaution of wearing masks and when they could take them off and know that they were safe?

This inquiry caused the garden slug specialist to remember his pets and the fact that they could smell cold boiled potatoes for a long distance. Perhaps, he said, these slugs could detect the presence of mustard gas.

It was known that mustard gas was injurious to human beings when there was one part of it present to four million parts of air. It had almost no odor and so might be present to that extent, might be burning out the lungs of the troops, and they would not know it, for they would not feel the effects until later. What was needed was some indicator of its presence before it got strong enough to do any harm to the troops, some detector that could give warning as to when it was time to put on the gas masks.

This government scientist put his garden slugs into gas chambers. He introduced mustard gas until there was one part of it present to twelve million parts of air. At this degree of dilution it was still harmless to human beings. But when the mustard got that thick the slugs began to act strangely. These odd animals have two eyes stuck out a half inch from their heads on the ends of horns or feelers—apparently a very handy place to carry eyes in order to see what is going on. Another peculiar thing about these eyes is the fact that they can be pulled in at will. The slugs began to pull in their eyes at the tip, as the end of a glove finger may be pulled in, when the mustard gas was one to twelve million. Thus was an indication to be had when it was present in that quantity.

The amount of the mustard in the chamber was increased until there was one part of it to eight million parts of air. At this stage the slugs began to excrete great quantities of their milky slime, a fact that could be readily seen by looking at them. The gas was burning them and they were trying to relieve themselves. Here was an indicator to show the presence of this amount of gas, still not enough to injure man. Then the density of the gas was increased to one part in four million, until it was dangerous to human beings. This was the point at which troops should put on their gas masks. When it got this strong the slug was writhing with agony. There

19

was no mistaking its suffering.   This was the final gas mask alarm.

When the Germans developed mustard gas they did so with a special purpose in mind, that of burning the lining out of the lungs of their enemies.   Mustard gas had in it certain chemical elements that, meeting with moisture, would unite with it.   When these united with water an entirely new chemical was made.   That chemical was hydrochloric acid.   Hydrochloric acid, when it comes in contact with the skin, burns it terribly.

The inside of the human lungs are always moist.   There is water always present.   When the mustard gas was breathed into the lungs it united with the moisture, and formed this acid which burned the lining of the lung, a membrane which is exceedingly thin and delicate.   It thus injured or killed its victim.

Now the outside of a slug is moist, as is the inside of the human lung.   As soon as the mustard gas appeared it united with the moisture on the outside of the slug and formed hydrochloric acid, which immediately began to burn the slug.   Hence its restlessness.   Hence its increased suffering as the strength of the gas was increased.   Hence its usefulness as a detector of mustard gas in the trenches.   The possibility of using it, however, would never have been considered had a scientific man not made investigations which, at the time, seemed entirely useless.

There was no difficulty in securing slugs for use at the front.   The common garden slug of America is a native of Europe and is very plentiful in northern France.   This is another unusual bit of information, the usefulness of which could not be foreseen. Two hours after the possibility of using slugs at the front was demonstrated in Washington a cable had been sent to France and England giving the facts and assuring those in command of the abundance of these slugs all about them.

When one attempts to find the place in the animal kingdom for so obscure a member of it as the garden slug he is likely to encounter some queer facts. An examination of the slug's body, for example, shows its very close similarity to that of the snail. The snail, one realizes, is a mollusk because it carries a shell on its back. It is a mollusk, as is the fresh water mussel, but of a different family. The fresh water mussel, we found, was a bivalve, that is, had two shells hinged together. The snail, on the contrary, has but a single shell which it carries about on its back and uses in emergencies as a place of safe retreat. It and the mussel are distant cousins.

A careful examination of the garden slug will show that it is a snail, that it has upon its back the remnant of what was once a shell. From its habit of hiding underneath some protecting board or stone or bit of rubbish, the slug has found that it can get along without a shell and so may relieve itself of the burden of carrying it about. Its shell has gradually disappeared. On its back, however, there still remains a bit of it as proof that it once existed. This shell is now of no importance to the slug, just as the wings of an ostrich, because of disuse, have come to be of no use to it for purposes of flying. Nevertheless, the slug is still a mollusk and is very closely related to the snail.

QUESTIONS

1. Did you ever see a garden slug? If not, go into any garden, or shed, or damp basement, and turn over the boards or look under vessels standing there and you are likely to find them.
2. What do the slugs do by night? What do they eat? Describe their method of building a road as they go. Did you ever see any of these roads?
3. Try, if possible, the experiment of putting ashes in the path of a slug. Tell what happens and why it happens. How may this knowledge be used in getting rid of garden slugs?
4. Describe the experiments of the scientist who kept slugs in his basement. What sense seemed very highly developed in them?
5. How may the sense of smell of the slug be tested on a lawn? In members of which branch of the animal kingdom does the sense of smell seem to be most highly developed?
6. Many strange contributions to victory are made in times of war. Tell the story of how it was found that the garden slug might be · used as a gas detector.
7. How was it that the slug was so sensitive to gas? How did it act under it? Why was it that mustard gas burned the lungs so terribly?
8. Do you know of other odd bits of information that have been found useful? How is it useful to know that plague is carried by rats? that poisonous snakes can be told by the pits in their heads? that toads may save the garden by eating the insects? that baby button clams steal rides on fishes? that house flies breed in manure piles?
9. The slug is a mollusk. Mollusks are animals that live in shells. How then can the slug, having no shell, be a mollusk? What is its closest relative?

What is a City?

Most people can agree that cities are places where large numbers of people live and work; they are hubs of government, commerce and transportation. But how best to define the geographical limits of a city is a matter of some debate. So far, no standardized international criteria exist for determining the boundaries of a city and often multiple different boundary definitions are available for any given city.

One type of definition, sometimes referred to as the "city proper", describes a city according to an administrative boundary. A second approach, termed the "urban agglomeration", considers the extent of the contiguous urban area, or built-up area, to delineate the city's boundaries. A third concept of the city, the "metropolitan area", defines its boundaries according to the degree of economic and social interconnectedness of nearby areas, identified by interlinked commerce or commuting patterns, for example.

The choice of how to define a city's boundaries is consequential for assessing the size of its population. In Toronto, Canada, for example, approximately 2.6 million people resided within the "city proper" according to the 2011 census, but the population of the surrounding "urban agglomeration" was almost twice as large, at 5.1 million, and the population of the "metropolitan area" was larger still, at 5.6 million.* Furthermore, rates of population growth differed across the three definitions. Between the 2006 and 2011 censuses, the population within Toronto's "city proper" grew at an average annual rate of 0.9 per cent, compared to 1.5 per cent for the "urban agglomeration" and 1.8 per cent for the "metropolitan area".

The 2014 revision of *World Urbanization Prospects* (WUP) endeavoured, wherever possible, given available data, to adhere to the "urban agglomeration" concept of cities. Very often, however, in order to compile a series of population estimates that was consistent for a city over time, the "city proper" or "metropolitan area" concepts were used instead. Of the 1,692 cities with at least 300,000 inhabitants in 2014 included in WUP, 55 per cent follow the "urban agglomeration" statistical concept, 35 per cent follow the "city proper" concept and the remaining 10 per cent refer to "metropolitan areas".

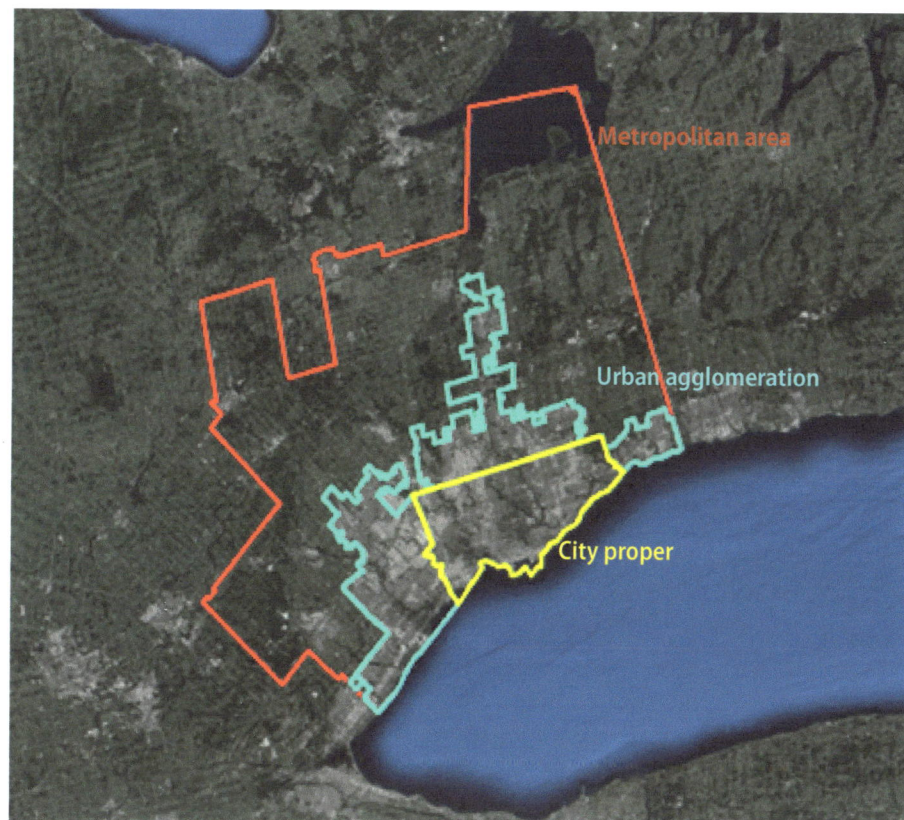

* The "city proper" described here corresponds to the Toronto "census subdivision – municipality" as defined in the 2011 Census of Canada; the "urban agglomeration" corresponds to the Toronto "population centre"; and the "metropolitan area" corresponds to the Toronto "census metropolitan area." Population data and boundaries are from Statistics Canada (http://www12.statcan.gc.ca/census-recensement/index-eng.cfm). Satellite image is from Google Imagery TerraMetrics 2016.

The world's cities are growing in both size and number

In 2016, there were 512 cities with at least 1 million inhabitants globally. By 2030, a projected 662 cities will have at least 1 million residents.

Cities with more than 10 million inhabitants are often termed "megacities". In 2016, there were 31 megacities globally and their number is projected to rise to 41 by 2030.

In 2016, 45 cities had populations between 5 and 10 million inhabitants. By 2030, 10 of these are projected to become megacities and the population of one (Saint Petersburg, Russian Federation) is expected to fall below 5 million. Projections indicate that 29 additional cities will cross the 5 million mark between 2016 and 2030, of which 15 are located in Asia and 10 in Africa. In 2030, 63 cities are projected to have between 5 and 10 million inhabitants.

An overwhelming majority of the world's cities have fewer than 5 million inhabitants. In 2016, there were 436 cities with between 1 and 5 million inhabitants and an additional 551 cities with between 500,000 and 1 million inhabitants. By 2030, the number of cities with 1 to 5 million inhabitants is projected to grow to 559 and 731 cities will have between 500,000 and 1 million inhabitants.

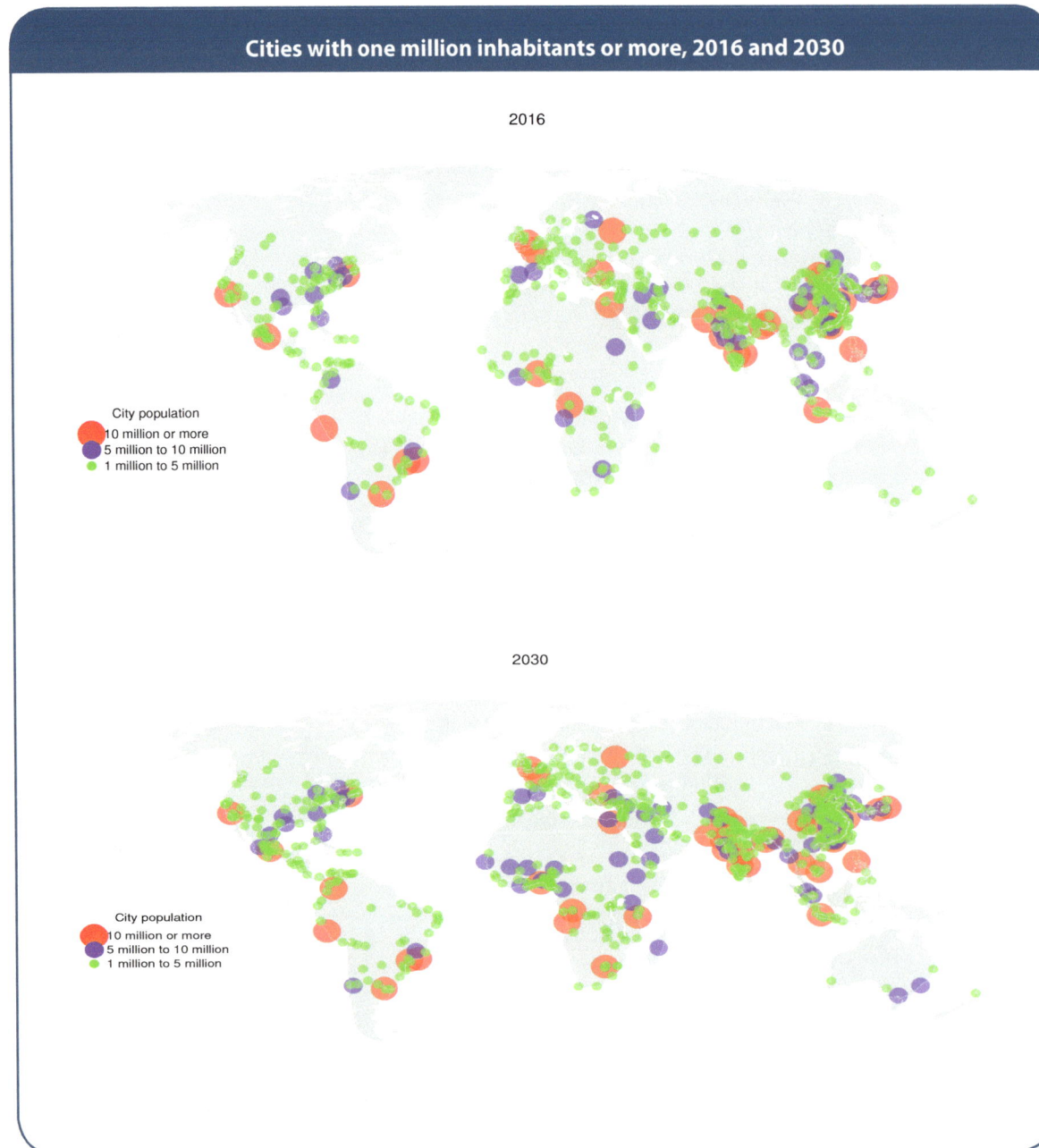

Cities with one million inhabitants or more, 2016 and 2030

2016

City population
10 million or more
5 million to 10 million
1 million to 5 million

2030

City population
10 million or more
5 million to 10 million
1 million to 5 million

One in five people worldwide lives in a city with more than 1 million inhabitants

In 2016, 1.7 billion people—23 per cent of the world's population—lived in a city with at least 1 million inhabitants. By 2030, a projected 27 per cent of people worldwide will be concentrated in cities with at least 1 million inhabitants.

Between 2016 and 2030, the population in all city size classes is projected to increase, while the rural population is projected to decline slightly. While rural areas were home to more than 45 per cent of the world's population in 2016, that proportion is expected to fall to 40 per cent by 2030.

A minority of people reside in megacities—500 million, representing 6.8 per cent of the global population in 2016. But, as these cities increase in both size and number, they will become home to a growing share of the population. By 2030, a projected 730 million people will live in cities with at least 10 million inhabitants, representing 8.7 per cent of people globally.

World's populationby size class of settlemwnt, 2016 and 2030

	2016			2030		
	Number of settlements	Population (billions)	Percentage of world population	Number of settlements	Population (billions)	Percentage of world population
Urban	..	4 034	54.5	..	5 058	60.0
10 million or more	31	500	6.8	41	730	8.7
5 to 10 million	45	308	4.2	63	434	5.2
1 to 5 million	436	861	11.6	558	1 128	13.4
500 000 to 1 million	551	380	5.1	731	509	6.0
Fewer than 500 000	..	1 985	26.8	..	2 257	26.8
Rural	..	3 371	45.5	..	3 367	40.0

World's population by size class of settlement, 1990-2030

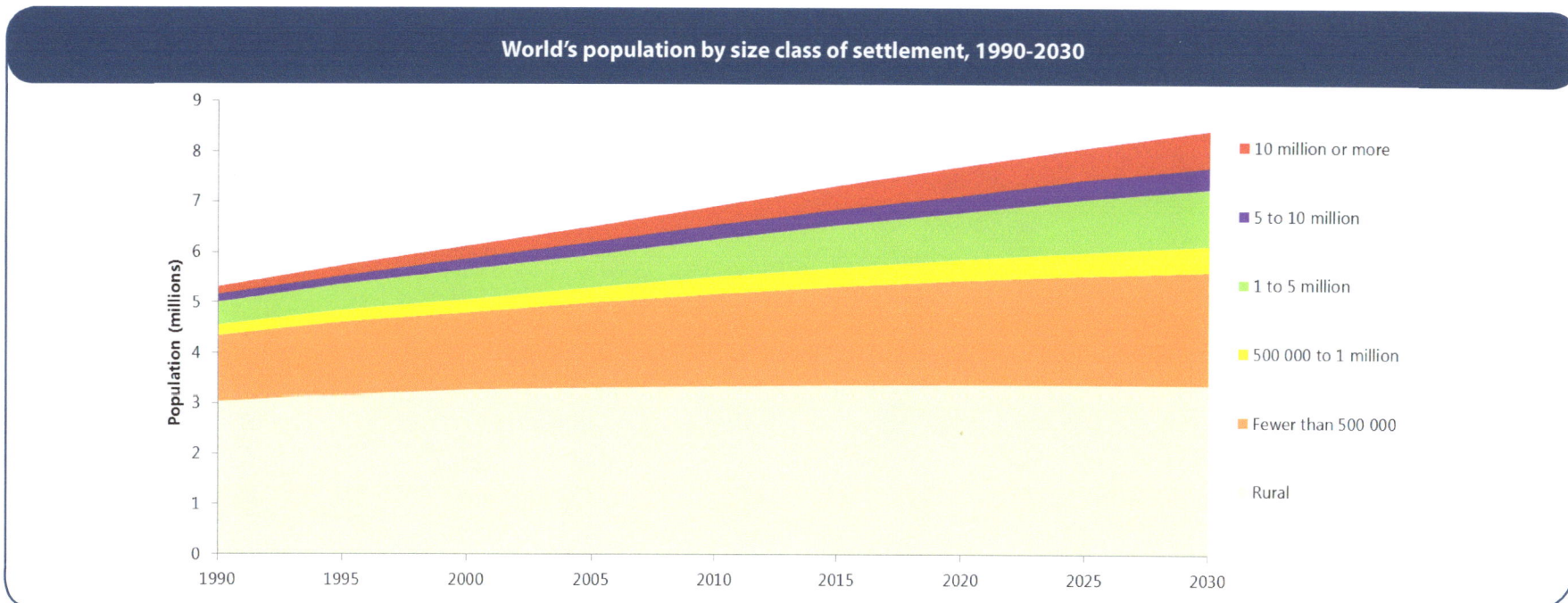

Most megacities are located in the global South

Of the world's 31 megacities (that is, cities with 10 million inhabitants or more) in 2016, 24 are located in the less developed regions or the "global South". China alone was home to six megacities in 2016, while India had five.

The 10 cities that are projected to become megacities between 2016 and 2030 are all located in developing countries. They include:

Lahore, Pakistan

Hyderabad, India

Bogotá, Colombia

Johannesburg, South Africa

Bangkok, Thailand

Dar es Salaam, Tanzania

Ahmanabad, India

Luanda, Angola

Ho Chi Minh City, Viet Nam

and Chungdu, China.

Despite a projected decline of nearly 1 million inhabitants, Tokyo is expected to remain the world's largest city in 2030, followed by Delhi, which is projected to add nearly 10 million people between 2016 and 2030.

Rank	City, Country	Population in 2016 (thousands)	City, Country	Population in 2030 (thousands)
1	Tokyo, Japan	38 140	Tokyo, Japan	37 190
2	Delhi, India	26 454	Delhi, India	36 060
3	Shanghai, China	24 484	Shanghai, China	30 751
4	Mumbai (Bombay), India	21 357	Mumbai (Bombay), India	27 797
5	São Paulo, Brazil	21 297	Beijing, China	27 706
6	Beijing, China	21 240	Dhaka, Bangladesh	27 374
7	Ciudad de México (Mexico City), Mexico	21 157	Karachi, Pakistan	24 838
8	Kinki M.M.A. (Osaka), Japan	20 337	Al-Qahirah (Cairo), Egypt	24 502
9	Al-Qahirah (Cairo), Egypt	19 128	Lagos, Nigeria	24 239
10	New York-Newark, USA	18 604	Ciudad de México (Mexico City), Mexico	23 865
11	Dhaka, Bangladesh	18 237	São Paulo, Brazil	23 444
12	Karachi, Pakistan	17 121	Kinshasa, Democratic Republic of the Congo	19 996
13	Buenos Aires, Argentina	15 334	Kinki M.M.A. (Osaka), Japan	19 976
14	Kolkata (Calcutta), India	14 980	New York-Newark, USA	19 885
15	Istanbul, Turkey	14 365	Kolkata (Calcutta), India	19 092
16	Chongqing, China	13 744	Guangzhou, Guangdong, China	17 574
17	Lagos, Nigeria	13 661	Chongqing, China	17 380
18	Manila, Philippines	13 131	Buenos Aires, Argentina	16 956
19	Guangzhou, Guangdong, China	13 070	Manila, Philippines	16 756
20	Rio de Janeiro, Brazil	12 981	Istanbul, Turkey	16 694
21	Los Angeles-Long Beach-Santa Ana, USA	12 317	Bangalore, India	14 762
22	Moskva (Moscow), Russian Federation	12 260	Tianjin, China	14 655
23	Kinshasa, Democratic Republic of the Congo	12 071	Rio de Janeiro, Brazil	14 174
24	Tianjin, China	11 558	Chennai (Madras), India	13 921
25	Paris, France	10 925	Jakarta, Indonesia	13 812
26	Shenzhen, China	10 828	Los Angeles-Long Beach-Santa Ana, USA	13 257
27	Jakarta, Indonesia	10 483	Lahore, Pakistan	13 033
28	Bangalore, India	10 456	Hyderabad, India	12 774
29	London, United Kingdom	10 434	Shenzhen, China	12 673
30	Chennai (Madras), India	10 163	Lima, Peru	12 221
31	Lima, Peru	10 072	Moskva (Moscow), Russian Federation	12 200
32			Bogotá, Colombia	11 966
33			Paris, France	11 803
34			Johannesburg, South Africa	11 573
35			Krung Thep (Bangkok), Thailand	11 528
36			London, United Kingdom	11 467
37			Dar es Salaam, United Republic of Tanzania	10 760
38			Ahmadabad, India	10 527
39			Luanda, Angola	10 429
40			Thành Pho Ho Chí Minh (Ho Chi Minh City), Viet Nam	10 200
41			Chengdu, China	10 104

The share of the population residing in cities is projected to increase in all regions

In Northern America, more than half of the population resided in cities with 500,000 inhabitants or more in 2016 and one in five people lived in a city of 5 million inhabitants or more. Latin America and the Caribbean is the region with the largest proportion of the population concentrated in megacities: of the total population of the region in 2016, 12.7 per cent resided in the five cities with 10 million inhabitants or more and the share in megacities is projected to rise to 14.3 per cent in 2030, as Bogotá crosses the 10 million threshold. In both Africa and Asia, more than half of the population lived in rural areas in 2016, but that share is declining. Between 2016 and 2030, the number of cities with 500,000 inhabitants or more is expected to grow by 80 per cent in Africa and by 30 per cent in Asia.

Population distribution by size class of settlement and region, 2016 and 2030

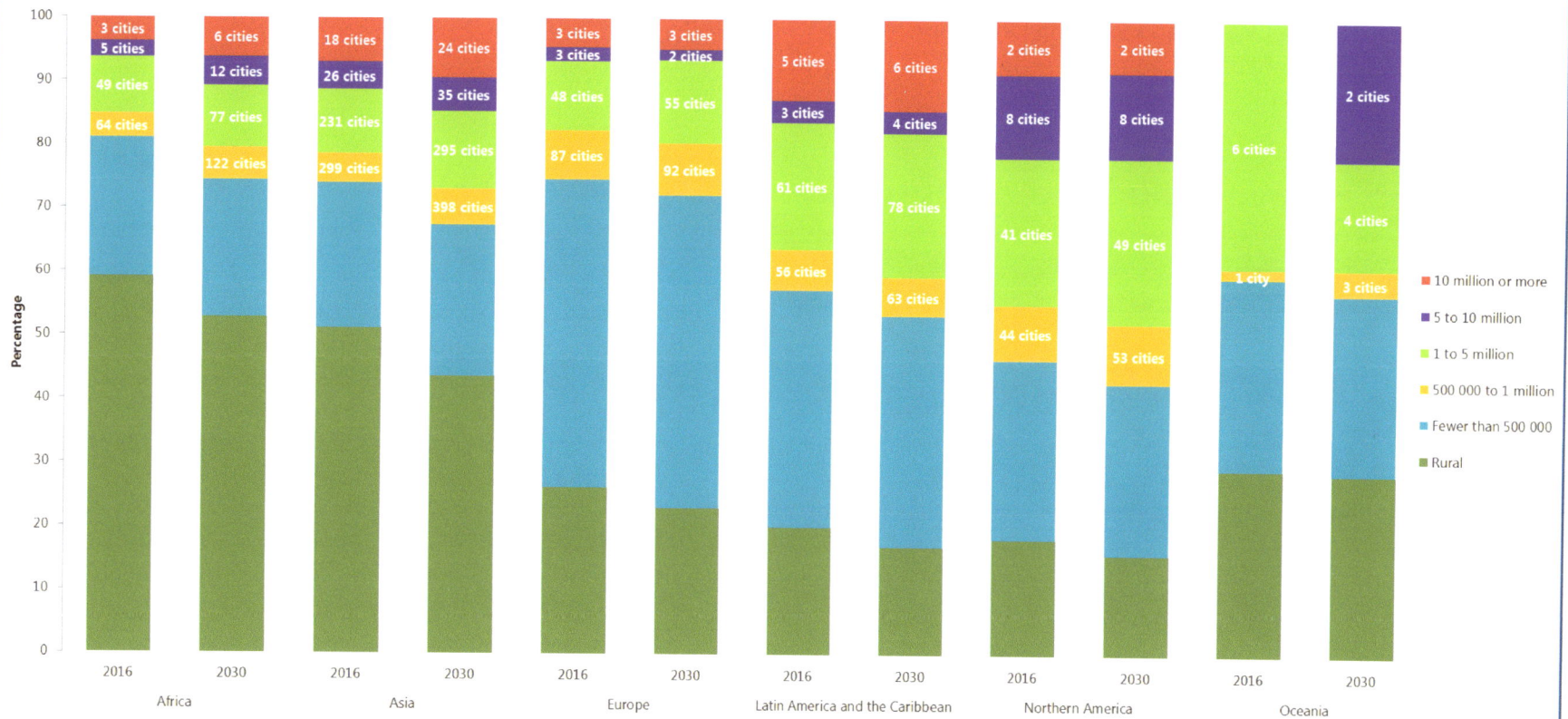

Legend:
- 10 million or more
- 5 to 10 million
- 1 to 5 million
- 500 000 to 1 million
- Fewer than 500 000
- Rural

Regions (2016 and 2030): Africa, Asia, Europe, Latin America and the Caribbean, Northern America, Oceania

Most of the world's fastest growing cities are located in Asia and Africa

Between 2000 and 2016, the world's cities with 500,000 inhabitants or more grew at an average annual rate of 2.4 per cent. However, 47 of these cities grew more than twice as fast, with average growth in excess of 6 per cent per year. Of these, 6 are located in Africa, 40 in Asia (20 in China alone), and 1 in Northern America. Among the fastest growing cities, 31 (nearly two thirds) have a long history of rapid population growth, with average annual growth rates above 6 per cent for the period 1980-2000 as well.

None of the 47 fastest growing cities had a population greater than 5 million in 2000, only 4 had between 1 and 5 million inhabitants, and 43 had fewer than 1 million inhabitants.

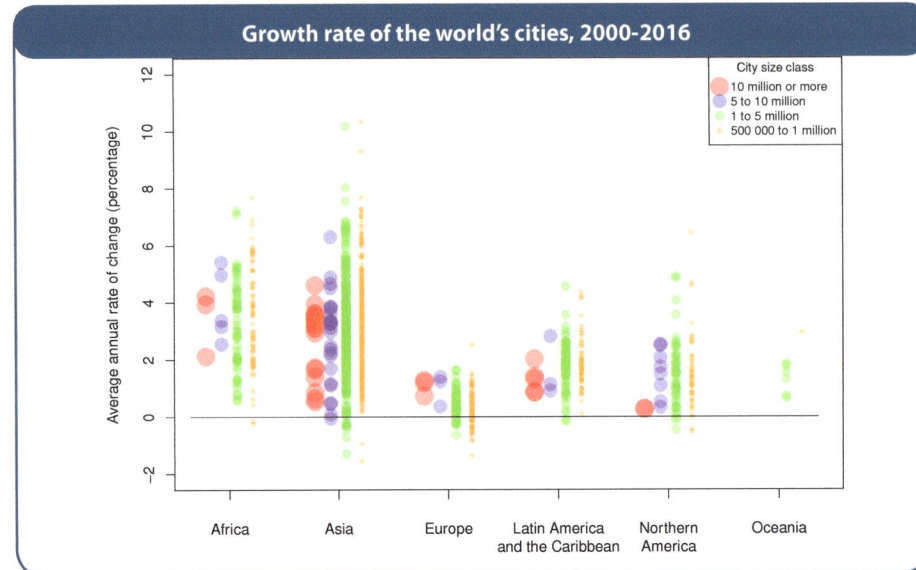

Growth rate of the world's cities, 2000-2016

City size class
- 10 million or more
- 5 to 10 million
- 1 to 5 million
- 500 000 to 1 million

Average annual rate of change (percentage)

Africa — Asia — Europe — Latin America and the Caribbean — Northern America — Oceania

Fifty-five cities have experienced population decline since 2000*

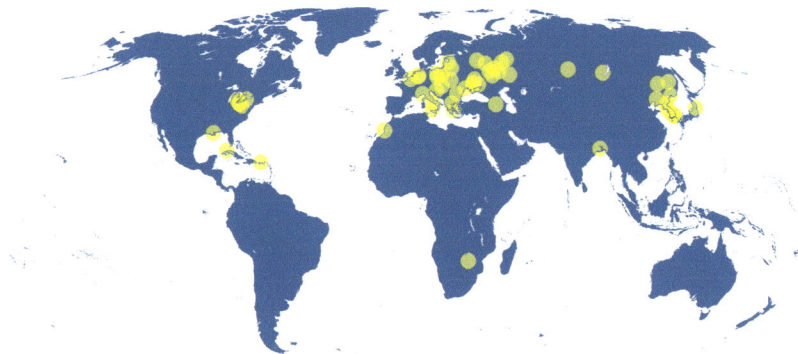

Cities where population declined between 2000 and 2016

Population decline in some cities occurred in response to a natural disaster. This has been the case in the United States city of New Orleans, which lost population after Hurricane Katrina in 2005, and in Sendai, Japan, following the 2011 earthquake and tsunami. Economic contraction has also contributed to population decline in places such as Buffalo and Detroit, concurrent with the loss of industry and jobs in those cities.

In most cases, however, city population decline has been associated not with crises, but rather with persistent low fertility rates, which have contributed to stagnating or declining total population sizes in some countries, particularly in Europe. The 55 cities with declining populations were home to 70 million people in 2016.

* Out of the 1,063 cities with 500,000 inhabitants or more in 2016.

In 28 countries or areas, more than 40 per cent of the urban population is concentrated in a single city of more than one million inhabitants

These "primate cities" include Hong Kong, Special Administrative Region (SAR) of China, with 7.4 million inhabitants in 2016, and Singapore, a city-state with 5.7 million inhabitants. An additional five cities concentrate more than 60 per cent of the urban population of their respective country or area, including Brazzaville (Congo), Kuwait City (Kuwait), Panama City (Panama), San Juan (Puerto Rico), and Ulaanbaatar (Mongolia).

For just over half of primate cities, the share of the urban population concentrated in the city has increased over time. The proportion of Mongolia's urban residents living in Ulaanbaatar, for example, rose from 56 per cent in 2000 to almost 66 per cent in 2016. The share of Georgia's urban population residing in Tbilisi increased from 44 per cent in 2000 to nearly 50 per cent in 2016.

Some primate cities are experiencing a decline in their share of the urban population. Lisbon, for example, held close to 48 per cent of the urban population of Portugal in 2000, but 43 per cent in 2016. The proportion of Guinea's urban dwellers residing in Conakry also declined from 45 per cent in 2000 to 42 per cent in 2016.

	Country or area	City	City population (thousands) 2000	2016	City population as a proportion of the urban population 2000	2016
1	China, Hong Kong SAR	Hong Kong	6 835	7 365	100	100
2	Singapore	Singapore	3 918	5 717	100	100
3	Kuwait	Al Kuwayt (Kuwait City)	1 300	2 874	69.5	79.4
4	Puerto Rico	San Juan	2 508	2 460	70.0	71.5
5	Mongolia	Ulaanbaatar	765	1 421	55.9	65.8
6	Panama	Ciudad de Panamá (Panama City)	1 216	1 708	64.0	63.1
7	Congo	Brazzaville	1 022	1 949	55.7	61.9
8	Liberia	Monrovia	836	1 305	65.2	56.5
9	Paraguay	Asunción	1 499	2 406	50.6	56.2
10	Armenia	Yerevan	1 111	1 040	55.9	55.6
11	Afghanistan	Kabul	2 401	4 842	54.8	54.5
12	Angola	Luanda	2 591	5 737	57.4	54.4
13	Senegal	Dakar	2 029	3 653	51.0	53.9
14	Cambodia	Phnum Pénh (Phnom Penh)	1 149	1 779	50.6	53.3
15	Uruguay	Montevideo	1 600	1 716	52.4	52.2
16	Burkina Faso	Ouagadougou	921	2 923	44.5	51.7
17	Egypt	Al-Qahirah (Cairo)	13 626	19 128	48.1	51.5
18	Lebanon	Bayrut (Beirut)	1 487	2 263	53.4	50.7
19	Georgia	Tbilisi	1 100	1 145	44.0	49.7
20	Israel	Tel Aviv-Yafo (Tel Aviv-Jaffa)	2 739	3 661	49.9	49.5
21	Somalia	Muqdisho (Mogadishu)	1 201	2 265	48.9	49.4
22	Azerbaijan	Baku	1 806	2 429	43.3	45.6
23	Portugal	Lisboa (Lisbon)	2 672	2 902	47.7	42.7
24	Côte d'Ivoire	Abidjan	3 028	5 020	43.1	42.0
25	Guinea	Conakry	1 221	1 989	45.0	41.7
26	Chad	N'Djaména	703	1 310	39.2	41.3
27	Peru	Lima	7 293	10 072	38.4	40.4
28	Chile	Santiago	5 658	6 544	42.5	40.4

Most cities are vulnerable to at least one type of natural disaster

Cities' risk of exposure to natural disasters

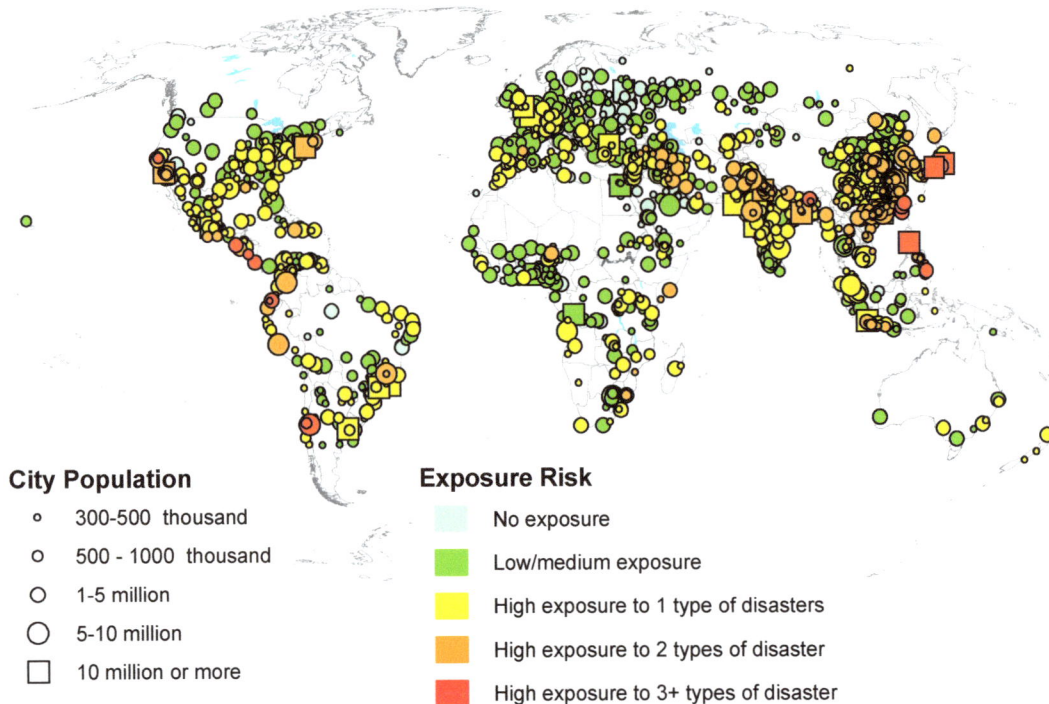

City Population

- ∘ 300-500 thousand
- ○ 500 - 1000 thousand
- ○ 1-5 million
- ○ 5-10 million
- □ 10 million or more

Exposure Risk

- No exposure
- Low/medium exposure
- High exposure to 1 type of disasters
- High exposure to 2 types of disaster
- High exposure to 3+ types of disaster

Of the 1,692 cities with at least 300,000 inhabitants in 2014, 944 (56 per cent) were at high risk of exposure to at least one of six types of natural disaster (cyclones, floods, droughts, earthquakes, landslides and volcano eruptions), based on evidence on the occurrence of natural disasters over the late twentieth century.* Taken together, cities facing high risk of exposure to a natural disaster were home to 1.4 billion people in 2014.

Around 15 per cent of cities—most located along coastlines—were at high risk of exposure to two or more types of natural disaster; 27 cities—including the megacities Tokyo, Osaka and Manila—faced high risk of exposure to three or more types of disaster.

* Results summarised here are from a 2015 United Nations technical paper that analysed city population estimates from the 2014 revision of World Urbanization Prospects together with spatial hotspot data on the risks of exposure and vulnerability to natural disasters produced by research institutes at Columbia University and the World Bank. The natural disaster data used in this analysis included historical information on the occurrence of six types of disaster, as well as their mortality and economic impacts. Data include cyclones occurring between 1980 and 2000; floods between 1985 and 2003; droughts between 1980 and 2000; earthquakes between 1976 and 2002; and landslides and volcano eruptions between 1979 and 2000. A city was classified as being at high risk of exposure or vulnerability to a particular type of natural disaster if its location intersected spatial grid cells ranking in the top three deciles of the global risk distribution in terms of the frequency of occurrence or mortality or economic impact of that specific type of natural disaster. For a complete description of the study and results see: Gu and others (2015) Risks of exposure and vulnerability to natural disasters at the city level: A global overview. Technical Paper No. 2015/2, Population Division, Department of Economic and Social Affairs, United Nations. Available from http://esa.un.org/unpd/wup/Publications/Files/WUP2014-TechnicalPaper-NaturalDisaster.pdf.

A majority of city dwellers live in cities that face high risk of disaster-related mortality or economic losses

Some 82 per cent of cities—home to 1.9 billion people in 2014—were located in areas that faced high risk of mortality associated with natural disasters. Similarly, 89 per cent of cities—home to 2.1 billion people in 2014—were located in areas that were highly vulnerable to economic losses associated with at least one of the six types of natural disaster.

On average, cities in the less developed regions were at higher risk of exposure to natural disasters and were more vulnerable to disaster-related economic losses and mortality than those in the more developed regions. Moreover, larger cities tended to be at higher risk of exposure to disasters and more vulnerable to disaster-related economic losses and mortality compared to smaller cities.

Floods were the most common type of natural disaster affecting cities, followed by droughts and cyclones. These three types of disaster were also the most devastating for city dwellers globally in terms of the mortality and economic losses they caused.

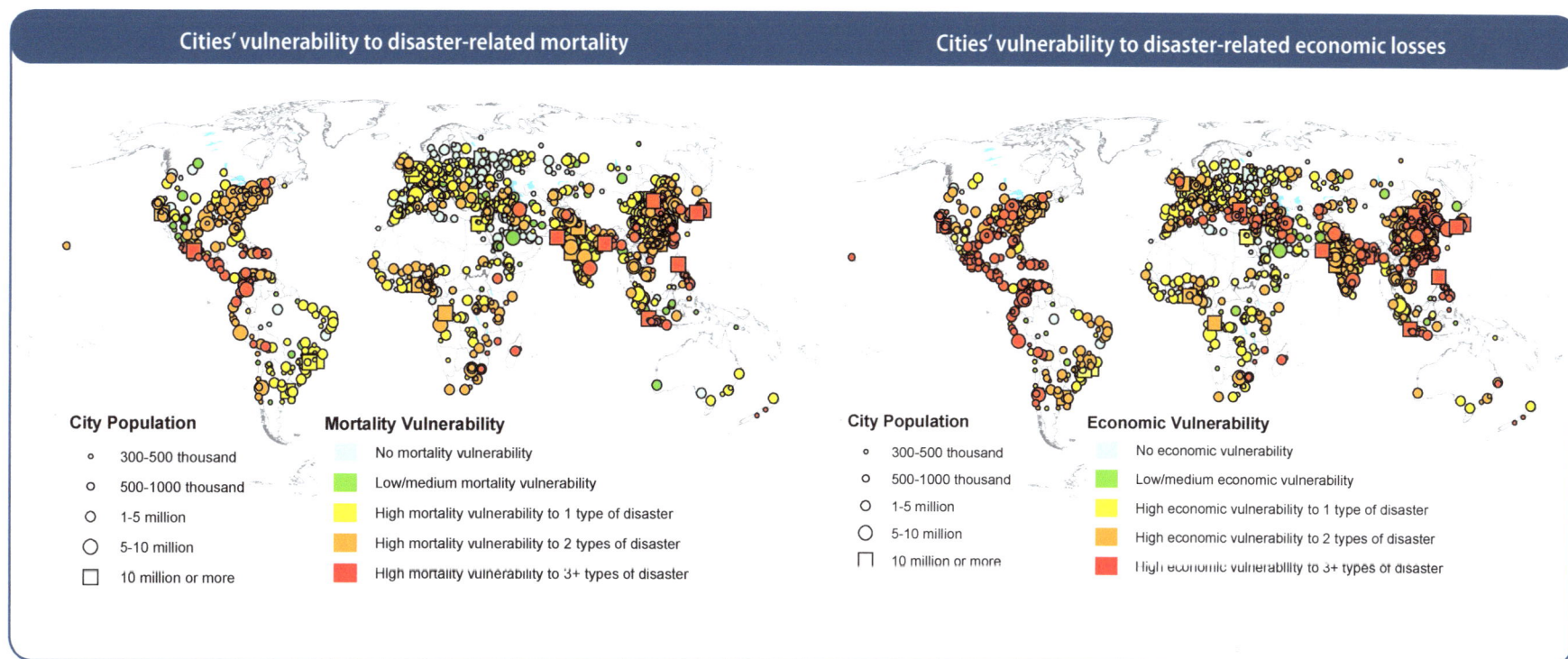

Cities' vulnerability to disaster-related mortality

Cities' vulnerability to disaster-related economic losses

City Population
- 300-500 thousand
- 500-1000 thousand
- 1-5 million
- 5-10 million
- 10 million or more

Mortality Vulnerability
- No mortality vulnerability
- Low/medium mortality vulnerability
- High mortality vulnerability to 1 type of disaster
- High mortality vulnerability to 2 types of disaster
- High mortality vulnerability to 3+ types of disaster

City Population
- 300-500 thousand
- 500-1000 thousand
- 1-5 million
- 5-10 million
- 10 million or more

Economic Vulnerability
- No economic vulnerability
- Low/medium economic vulnerability
- High economic vulnerability to 1 type of disaster
- High economic vulnerability to 2 types of disaster
- High economic vulnerability to 3+ types of disaster

The boundaries and names shown and the designations used on these maps do not imply official endorsement or acceptance by the United Nations. Dotted line represents approximately the Line of Control in Jammu and Kashmir agreed upon by India and Pakistan. The final status of Jammu and Kashmir has not yet been agreed upon by the parties. Final boundary between the Republic of Sudan and the Republic of South Sudan has not yet been determined.

Country or area	City	Statistical concept	City population (thousands)			Average annual rate of change (percentage)		City population as a proportion of the country or area's total or urban population in 2016 (percentage)	
			2000	2016	2030	2000-2016	2016-2030	Total population	Urban population
Africa									
Eastern Africa									
Ethiopia	Addis Ababa	City Proper	2 377	3 316	5 851	2.1	4.1	3.3	16.4
Kenya	Mombasa	City Proper	683	1 141	1 973	3.2	3.9	2.4	9.1
Kenya	Nairobi	City Proper	2 214	4 070	7 140	3.8	4.0	8.5	32.6
Madagascar	Antananarivo	Urban Agglomeration	1 361	2 739	5 073	4.4	4.4	11.0	30.8
Mozambique	Maputo	Urban Agglomeration	1 019	1 203	1 893	1.0	3.2	4.3	13.3
Rwanda	Kigali	Urban Agglomeration	578	1 293	2 268	5.0	4.0	10.1	34.0
Somalia	Muqdisho (Mogadishu)	Urban Agglomeration	1 201	2 265	4 176	4.0	4.4	19.8	49.4
Uganda	Kampala	Urban Agglomeration	1 097	2 012	3 939	3.8	4.8	4.9	29.5
United Republic of Tanzania	Dar es Salaam	Urban Agglomeration	2 272	5 409	10 760	5.4	4.9	10.0	31.1
Zambia	Lusaka	Urban Agglomeration	1 073	2 285	4 365	4.7	4.6	14.3	34.4
Zimbabwe	Harare	City Proper	1 379	1 511	2 048	0.6	2.2	9.8	30.2
Middle Africa									
Angola	Huambo	Urban Agglomeration	578	1 337	2 537	5.2	4.6	5.7	12.7
Angola	Luanda	Urban Agglomeration	2 591	5 737	10 429	5.0	4.3	24.4	54.4
Cameroon	Douala	Urban Agglomeration	1 490	3 051	4 774	4.5	3.2	12.7	23.2
Cameroon	Yaoundé	Urban Agglomeration	1 351	3 204	5 158	5.4	3.4	13.4	24.3
Chad	N'Djaména	City Proper	703	1 310	2 347	3.9	4.2	9.4	41.3
Congo	Brazzaville	City Proper	1 022	1 949	2 979	4.0	3.0	40.7	61.9
Dem. Republic of the Congo	Kananga	City Proper	604	1 219	2 053	4.4	3.7	1.7	3.9
Dem. Republic of the Congo	Kinshasa	Urban Agglomeration	6 140	12 071	19 996	4.2	3.6	16.5	38.4
Dem. Republic of the Congo	Kisangani	City Proper	586	1 079	1 786	3.8	3.6	1.5	3.4
Dem. Republic of the Congo	Lubumbashi	City Proper	1 089	2 097	3 489	4.1	3.6	2.9	6.7
Dem. Republic of the Congo	Mbuji-Mayi	City Proper	1 011	2 097	3 541	4.6	3.7	2.9	6.7
Northern Africa									
Algeria	El Djazaïr (Algiers)	Urban Agglomeration	2 141	2 632	3 405	1.3	1.8	6.4	8.9
Egypt	Al-Iskandariyah (Alexandria)	City Proper	3 546	4 863	6 313	2.0	1.9	5.7	13.1
Egypt	Al-Qahirah (Cairo)	Metropolitan area	13 626	19 128	24 502	2.1	1.8	22.2	51.5
Libya	Tarabulus (Tripoli)	City Proper	1 022	1 128	1 333	0.6	1.2	17.6	22.4

Country or area	City	Statistical Concepts	City population (thousands)			Average annual rate of change (percentage)		City population as a proportion of the country or area's total or urban population in 2016 (percentage)	
			2000	2016	2030	2000-2016	2016-2030	Total population	Urban population
Morocco	Dar-el-Beida (Casablanca)	Urban Agglomeration	3 134	3 544	4 361	0.8	1.5	10.3	17.0
Morocco	Fès	Urban Agglomeration	868	1 197	1 559	2.0	1.9	3.5	5.7
Morocco	Marrakech	City Proper	751	1 168	1 572	2.8	2.1	3.4	5.6
Morocco	Rabat	Urban Agglomeration	1 507	2 004	2 574	1.8	1.8	5.8	9.6
Morocco	Tanger	Urban Agglomeration	591	1 016	1 391	3.4	2.2	3.0	4.9
Sudan	Al-Khartum (Khartoum)	Urban Agglomeration	3 505	5 265	8 158	2.5	3.1	13.0	38.2
Tunisia	Tunis	Urban Agglomeration	1 769	2 010	2 347	0.8	1.1	17.7	26.4
Southern Africa									
South Africa	Cape Town	Urban Agglomeration	2 715	3 698	4 322	1.9	1.1	6.9	10.5
South Africa	Durban	Urban Agglomeration	2 370	2 914	3 349	1.3	1.0	5.4	8.3
South Africa	Johannesburg	Metropolitan area	5 605	9 616	11 573	3.4	1.3	17.9	27.4
South Africa	Port Elizabeth	Metropolitan area	958	1 186	1 390	1.3	1.1	2.2	3.4
South Africa	Pretoria	Urban Agglomeration	1 084	2 125	2 701	4.2	1.7	3.9	6.0
South Africa	Vereeniging	Urban Agglomeration	897	1 164	1 370	1.6	1.2	2.2	3.3
Western Africa									
Burkina Faso	Ouagadougou	City Proper	921	2 923	5 854	7.2	5.0	15.9	51.7
Côte d'Ivoire	Abidjan	Urban Agglomeration	3 028	5 020	7 773	3.2	3.1	23.0	42.0
Ghana	Accra	Metropolitan area	1 668	2 316	3 262	2.0	2.4	8.4	15.4
Ghana	Kumasi	Metropolitan area	1 187	2 718	4 215	5.2	3.1	9.9	18.1
Guinea	Conakry	Urban Agglomeration	1 221	1 989	3 134	3.0	3.2	15.7	41.7
Liberia	Monrovia	Urban Agglomeration	836	1 305	2 032	2.8	3.2	28.3	56.5
Mali	Bamako	City Proper	1 142	2 651	5 231	5.3	4.9	15.8	38.8
Niger	Niamey	City Proper	696	1 125	2 363	3.0	5.3	5.6	29.5
Nigeria	Abuja	Urban Agglomeration	833	2 586	4 913	7.1	4.6	1.4	2.8
Nigeria	Benin City	Urban Agglomeration	975	1 543	2 667	2.9	3.9	0.8	1.7
Nigeria	Ibadan	Urban Agglomeration	2 236	3 243	5 499	2.3	3.8	1.7	3.5
Nigeria	Kaduna	City Proper	881	1 064	1 748	1.2	3.5	0.6	1.2
Nigeria	Kano	Urban Agglomeration	2 602	3 676	6 198	2.2	3.7	1.9	4.0
Nigeria	Lagos	Urban Agglomeration	7 281	13 661	24 239	3.9	4.1	7.2	14.9
Nigeria	Onitsha	Urban Agglomeration	533	1 165	2 147	4.9	4.4	0.6	1.3

Country or area	City	Statistical Concepts	City population (thousands)			Average annual rate of change (percentage)		City population as a proportion of the country or area's total or urban population in 2016 (percentage)	
			2000	2016	2030	2000-2016	2016-2030	Total population	Urban population
Nigeria	Port Harcourt	Urban Agglomeration	1 091	2 465	4 562	5.1	4.4	1.3	2.7
Senegal	Dakar	Urban Agglomeration	2 029	3 653	6 046	3.7	3.6	23.7	53.9
Sierra Leone	Freetown	City Proper	690	1 029	1 470	2.5	2.5	16.0	39.7
Asia									
Eastern Asia									
China	Anshan	Urban Agglomeration	1 384	1 570	1 886	0.8	1.3	0.1	0.2
China	Anyang	Urban Agglomeration	548	1 191	1 633	4.9	2.2	0.1	0.1
China	Baoding	Urban Agglomeration	884	1 120	1 377	1.5	1.5	0.1	0.1
China	Baoji	Urban Agglomeration	638	1 028	1 329	3.0	1.8	0.1	0.1
China	Baotou	Urban Agglomeration	1 406	1 996	2 495	2.2	1.6	0.1	0.2
China	Beijing	Urban Agglomeration	10 162	21 240	27 706	4.6	1.9	1.5	2.7
China	Benxi	Urban Agglomeration	857	1 085	1 338	1.5	1.5	0.1	0.1
China	Changchun	Urban Agglomeration	2 730	3 835	4 742	2.1	1.5	0.3	0.5
China	Changsha	Urban Agglomeration	2 182	3 882	5 013	3.6	1.8	0.3	0.5
China	Changzhou, Jiangsu	Urban Agglomeration	1 478	2 653	3 381	3.7	1.7	0.2	0.3
China	Chaozhou	Urban Agglomeration	691	1 349	1 650	4.2	1.4	0.1	0.2
China	Chengdu	Urban Agglomeration	4 222	7 820	10 104	3.9	1.8	0.6	1.0
China	Chifeng	Urban Agglomeration	677	1 043	1 333	2.7	1.8	0.1	0.1
China	Chongqing	Urban Agglomeration	7 863	13 744	17 380	3.5	1.7	1.0	1.7
China	Cixi	Urban Agglomeration	650	1 356	1 829	4.6	2.1	0.1	0.2
China	Dalian	Urban Agglomeration	2 833	4 612	5 851	3.0	1.7	0.3	0.6
China	Daqing	Urban Agglomeration	1 082	1 661	2 113	2.7	1.7	0.1	0.2
China	Datong	Urban Agglomeration	1 049	1 567	1 988	2.5	1.7	0.1	0.2
China	Dongguan	Urban Agglomeration	3 631	7 469	8 701	4.5	1.1	0.5	0.9
China	Foshan	Urban Agglomeration	3 832	7 089	8 353	3.8	1.2	0.5	0.9
China	Fushun, Liaoning	Urban Agglomeration	1 358	1 295	1 508	-0.3	1.1	0.1	0.2
China	Fuzhou, Fujian	Urban Agglomeration	2 009	3 380	4 335	3.3	1.8	0.2	0.4
China	Guangzhou, Guangdong	Urban Agglomeration	7 330	13 070	17 574	3.6	2.1	0.9	1.6
China	Guilin	Urban Agglomeration	805	1 056	1 310	1.7	1.5	0.1	0.1
China	Guiyang	Urban Agglomeration	1 860	2 944	3 734	2.9	1.7	0.2	0.4
China	Haerbin	Urban Agglomeration	3 888	5 565	6 860	2.2	1.5	0.4	0.7

Country or area	City	Statistical Concepts	City population (thousands)			Average annual rate of change (percentage)		City population as a proportion of the country or area's total or urban population in 2016 (percentage)	
			2000	2016	2030	2000-2016	2016-2030	Total population	Urban population
China	Qinhuangdao	Urban Agglomeration	702	1 139	1 470	3.0	1.8	0.1	0.1
China	Qiqihaer	Urban Agglomeration	1 188	1 469	1 798	1.3	1.4	0.1	0.2
China	Quanzhou	Urban Agglomeration	728	1 447	1 928	4.3	2.0	0.1	0.2
China	Rizhao	Urban Agglomeration	613	1 096	1 441	3.6	2.0	0.1	0.1
China	Ruian	Urban Agglomeration	555	1 005	1 326	3.7	2.0	0.1	0.1
China	Shanghai	City Proper	13 959	24 484	30 751	3.5	1.6	1.7	3.1
China	Shantou	Urban Agglomeration	2 931	4 011	4 899	2.0	1.4	0.3	0.5
China	Shaoxing	Urban Agglomeration	1 124	2 151	2 841	4.1	2.0	0.2	0.3
China	Shenyang	Urban Agglomeration	4 562	6 438	7 911	2.2	1.5	0.5	0.8
China	Shenzhen	Urban Agglomeration	6 550	10 828	12 673	3.1	1.1	0.8	1.4
China	Shijiazhuang	Urban Agglomeration	1 914	3 370	4 362	3.5	1.8	0.2	0.4
China	Suqian	Urban Agglomeration	397	1 111	1 591	6.4	2.6	0.1	0.1
China	Suzhou, Jiangsu	Urban Agglomeration	2 112	5 788	8 098	6.3	2.4	0.4	0.7
China	Taian, Shandong	Urban Agglomeration	910	1 239	1 541	1.9	1.6	0.1	0.2
China	Taichung	Urban Agglomeration	978	1 241	1 528	1.5	1.5	0.1	0.2
China	Taipei	Urban Agglomeration	2 630	2 669	3 116	0.1	1.1	0.2	0.3
China	Taiyuan, Shanxi	Urban Agglomeration	2 503	3 549	4 396	2.2	1.5	0.3	0.4
China	Taizhou, Jiangsu	Urban Agglomeration	784	1 211	1 541	2.7	1.7	0.1	0.2
China	Taizhou, Zhejiang	City Proper	992	1 695	2 189	3.4	1.8	0.1	0.2
China	Tangshan, Hebei	Urban Agglomeration	1 418	2 853	3 801	4.4	2.0	0.2	0.4
China	Tianjin	Urban Agglomeration	6 670	11 558	14 655	3.4	1.7	0.8	1.4
China	Ürümqi (Wulumqi)	Urban Agglomeration	1 807	3 639	4 831	4.4	2.0	0.3	0.5
China	Weifang	Urban Agglomeration	1 235	2 269	2 973	3.8	1.9	0.2	0.3
China	Wenzhou	Urban Agglomeration	1 565	3 319	4 336	4.7	1.9	0.2	0.4
China	Wuhan	Urban Agglomeration	6 638	7 979	9 442	1.2	1.2	0.6	1.0
China	Wuhu, Anhui	Urban Agglomeration	634	1 495	2 074	5.4	2.3	0.1	0.2
China	Wuxi, Jiangsu	Urban Agglomeration	1 835	3 109	3 862	3.3	1.5	0.2	0.4
China	Xi'an, Shaanxi	Urban Agglomeration	3 690	6 220	7 904	3.3	1.7	0.4	0.8
China	Xiamen	Urban Agglomeration	1 416	4 738	6 911	7.5	2.7	0.3	0.6
China	Xiangtan, Hunan	Urban Agglomeration	698	1 032	1 311	2.4	1.7	0.1	0.1
China	Xiangyang	Urban Agglomeration	1 202	1 554	1 907	1.6	1.5	0.1	0.2

Country or area	City	Statistical Concepts	City population (thousands)			Average annual rate of change (percentage)		City population as a proportion of the country or area's total or urban population in 2016 (percentage)	
			2000	2016	2030	2000-2016	2016-2030	Total population	Urban population
China	Xining	Urban Agglomeration	844	1 359	1 750	3.0	1.8	0.1	0.2
China	Xinxiang	Urban Agglomeration	762	1 006	1 249	1.7	1.5	0.1	0.1
China	Xuzhou	Urban Agglomeration	1 367	1 956	2 445	2.2	1.6	0.1	0.2
China	Yancheng, Jiangsu	Urban Agglomeration	671	1 502	2 059	5.0	2.3	0.1	0.2
China	Yangzhou	Urban Agglomeration	1 217	1 803	2 269	2.5	1.6	0.1	0.2
China	Yantai	Urban Agglomeration	1 218	2 182	2 838	3.6	1.9	0.2	0.3
China	Yichang	Urban Agglomeration	692	1 310	1 743	4.0	2.0	0.1	0.2
China	Yinchuan	Urban Agglomeration	571	1 698	2 461	6.8	2.7	0.1	0.2
China	Yingkou	Urban Agglomeration	624	1 057	1 380	3.3	1.9	0.1	0.1
China	Yiwu	Urban Agglomeration	532	1 124	1 520	4.7	2.2	0.1	0.1
China	Zaozhuang	Urban Agglomeration	853	1 038	1 264	1.2	1.4	0.1	0.1
China	Zhanjiang	Urban Agglomeration	818	1 172	1 476	2.2	1.7	0.1	0.1
China	Zhengzhou	Urban Agglomeration	2 438	4 539	5 900	3.9	1.9	0.3	0.6
China	Zhenjiang, Jiangsu	Urban Agglomeration	743	1 070	1 349	2.3	1.7	0.1	0.1
China	Zhongshan	Urban Agglomeration	1 376	3 908	5 518	6.5	2.5	0.3	0.5
China	Zhuhai	Urban Agglomeration	1 004	1 578	2 003	2.8	1.7	0.1	0.2
China	Zhuzhou	Urban Agglomeration	819	1 100	1 369	1.8	1.6	0.1	0.1
China	Zibo	Urban Agglomeration	1 874	2 465	3 015	1.7	1.4	0.2	0.3
China, Hong Kong SAR	Hong Kong	Urban Agglomeration	6 835	7 365	7 885	0.5	0.5	100.0	100.0
Dem. People's Rep. of Korea	P'yongyang	City Proper	2 777	2 872	3 277	0.2	0.9	11.4	18.6
Japan	Chukyo M.M.A. (Nagoya)	Metropolitan area	8 740	9 434	9 304	0.5	-0.1	7.5	7.9
Japan	Hiroshima	Metropolitan area	2 044	2 180	2 213	0.4	0.1	1.7	1.8
Japan	Kinki M.M.A. (Osaka)	Metropolitan area	18 660	20 337	19 976	0.5	-0.1	16.1	17.1
Japan	Kitakyushu-Fukuoka M.M.A.	Metropolitan area	5 421	5 494	5 355	0.1	-0.2	4.3	4.6
Japan	Sapporo	Metropolitan area	2 508	2 564	2 542	0.1	-0.1	2.0	2.2
Japan	Sendai	Metropolitan area	2 184	2 071	2 012	-0.3	-0.2	1.6	1.7
Japan	Shizuoka-Hamamatsu M.M.A.	Metropolitan area	1 217	3 493	3 934	6.6	0.8	2.8	2.9
Japan	Tokyo	Metropolitan area	34 450	38 140	37 190	0.6	-0.2	30.1	32.1
Mongolia	Ulaanbaatar	City Proper	765	1 421	1 850	3.9	1.9	47.9	65.8
Republic of Korea	Busan	City Proper	3 594	3 200	3 264	-0.7	0.1	6.4	7.8
Republic of Korea	Changwon	Urban Agglomeration	1 077	1 036	1 090	-0.2	0.4	2.1	2.5

Country or area	City	Statistical Concepts	City population (thousands)			Average annual rate of change (percentage)		City population as a proportion of the country or area's total or urban population in 2016 (percentage)	
			2000	2016	2030	2000-2016	2016-2030	Total population	Urban population
Republic of Korea	Daegu	City Proper	2 323	2 241	2 328	-0.2	0.3	4.5	5.4
Republic of Korea	Daejon	City Proper	1 354	1 578	1 711	1.0	0.6	3.2	3.8
Republic of Korea	Gwangju	City Proper	1 343	1 550	1 682	0.9	0.6	3.1	3.8
Republic of Korea	Incheon	City Proper	2 371	2 711	2 919	0.8	0.5	5.4	6.6
Republic of Korea	Seoul	City Proper	9 878	9 779	9 960	-0.1	0.1	19.6	23.7
Republic of Korea	Suweon	City Proper	932	1 106	1 195	1.1	0.6	2.2	2.7
Republic of Korea	Yongin	City Proper	376	1 090	1 325	6.7	1.4	2.2	2.6
South-Central Asia									
Central Asia									
Kazakhstan	Almaty	City Proper	1 160	1 535	1 730	1.8	0.9	9.1	17.0
Uzbekistan	Tashkent	City Proper	2 135	2 264	2 839	0.4	1.6	7.5	20.6
Southern Asia									
Afghanistan	Kabul	City Proper	2 401	4 842	8 280	4.4	3.8	14.8	54.5
Bangladesh	Chittagong	Metropolitan area	3 308	4 640	6 719	2.1	2.6	2.9	8.2
Bangladesh	Dhaka	Metropolitan area	10 285	18 237	27 374	3.6	2.9	11.2	32.1
Bangladesh	Khulna	Metropolitan area	1 247	1 013	1 336	-1.3	2.0	0.6	1.8
India	Agra	Urban Agglomeration	1 293	2 017	2 793	2.8	2.3	0.2	0.5
India	Ahmadabad	Urban Agglomeration	4 427	7 571	10 527	3.4	2.4	0.6	1.8
India	Aligarh	Urban Agglomeration	653	1 067	1 502	3.1	2.4	0.1	0.2
India	Allahabad	Urban Agglomeration	1 035	1 313	1 755	1.5	2.1	0.1	0.3
India	Amritsar	Urban Agglomeration	990	1 283	1 721	1.6	2.1	0.1	0.3
India	Asansol	Urban Agglomeration	1 065	1 330	1 768	1.4	2.0	0.1	0.3
India	Aurangabad	Urban Agglomeration	868	1 380	1 925	2.9	2.4	0.1	0.3
India	Bangalore	Urban Agglomeration	5 567	10 456	14 762	3.9	2.5	0.8	2.4
India	Bareilly	Urban Agglomeration	722	1 141	1 598	2.9	2.4	0.1	0.3
India	Bhopal	Urban Agglomeration	1 426	2 151	2 959	2.6	2.3	0.2	0.5
India	Bhubaneswar	Urban Agglomeration	637	1 026	1 439	3.0	2.4	0.1	0.2
India	Chandigarh	Urban Agglomeration	791	1 159	1 595	2.4	2.3	0.1	0.3
India	Chennai (Madras)	Urban Agglomeration	6 353	10 163	13 921	2.9	2.2	0.8	2.4
India	Coimbatore	Urban Agglomeration	1 420	2 641	3 782	3.9	2.6	0.2	0.6

Country or area	City	Statistical Concepts	City population (thousands)			Average annual rate of change (percentage)		City population as a proportion of the country or area's total or urban population in 2016 (percentage)	
			2000	2016	2030	2000-2016	2016-2030	Total population	Urban population
India	Delhi	Urban Agglomeration	15 732	26 454	36 060	3.2	2.2	2.0	6.2
India	Dhanbad	Urban Agglomeration	1 046	1 269	1 680	1.2	2.0	0.1	0.3
India	Durg-Bhilainagar	Urban Agglomeration	905	1 144	1 529	1.5	2.1	0.1	0.3
India	Guwahati (Gauhati)	Urban Agglomeration	797	1 059	1 430	1.8	2.1	0.1	0.2
India	Gwalior	Urban Agglomeration	855	1 248	1 718	2.4	2.3	0.1	0.3
India	Hubli-Dharwad	City Proper	776	1 037	1 405	1.8	2.2	0.1	0.2
India	Hyderabad	Urban Agglomeration	5 445	9 218	12 774	3.3	2.3	0.7	2.1
India	Indore	Urban Agglomeration	1 597	2 503	3 459	2.8	2.3	0.2	0.6
India	Jabalpur	Urban Agglomeration	1 100	1 352	1 795	1.3	2.0	0.1	0.3
India	Jaipur	Urban Agglomeration	2 259	3 549	4 885	2.8	2.3	0.3	0.8
India	Jamshedpur	Urban Agglomeration	1 081	1 477	2 000	2.0	2.2	0.1	0.3
India	Jodhpur	Urban Agglomeration	842	1 318	1 837	2.8	2.4	0.1	0.3
India	Kannur	Urban Agglomeration	837	2 278	3 513	6.3	3.1	0.2	0.5
India	Kanpur	Urban Agglomeration	2 641	3 044	3 950	0.9	1.9	0.2	0.7
India	Kochi (Cochin)	Urban Agglomeration	1 523	2 484	3 465	3.1	2.4	0.2	0.6
India	Kolkata (Calcutta)	Urban Agglomeration	13 058	14 980	19 092	0.9	1.7	1.2	3.5
India	Kollam	Urban Agglomeration	611	1 482	2 246	5.5	3.0	0.1	0.3
India	Kota	Urban Agglomeration	692	1 200	1 710	3.4	2.5	0.1	0.3
India	Kozhikode (Calicut)	Urban Agglomeration	1 237	2 582	3 776	4.6	2.7	0.2	0.6
India	Lucknow	Urban Agglomeration	2 221	3 295	4 493	2.5	2.2	0.3	0.8
India	Ludhiana	Urban Agglomeration	1 368	1 739	2 316	1.5	2.0	0.1	0.4
India	Madurai	Urban Agglomeration	1 187	1 623	2 201	2.0	2.2	0.1	0.4
India	Malappuram	Urban Agglomeration	875	2 342	3 600	6.2	3.1	0.2	0.5
India	Meerut	Urban Agglomeration	1 143	1 579	2 140	2.0	2.2	0.1	0.4
India	Moradabad	Urban Agglomeration	626	1 054	1 493	3.3	2.5	0.1	0.2
India	Mumbai (Bombay)	Urban Agglomeration	16 367	21 357	27 797	1.7	1.9	1.6	5.0
India	Mysore	Urban Agglomeration	776	1 105	1 515	2.2	2.3	0.1	0.3
India	Nagpur	Urban Agglomeration	2 089	2 715	3 616	1.6	2.0	0.2	0.6
India	Nashik	Urban Agglomeration	1 117	1 829	2 556	3.1	2.4	0.1	0.4
India	Patna	Urban Agglomeration	1 658	2 247	3 016	1.9	2.1	0.2	0.5
India	Pune (Poona)	Urban Agglomeration	3 655	5 882	8 091	3.0	2.3	0.5	1.4

Country or area	City	Statistical Concepts	City population (thousands)			Average annual rate of change (percentage)		City population as a proportion of the country or area's total or urban population in 2016 (percentage)	
			2000	2016	2030	2000-2016	2016-2030	Total population	Urban population
India	Raipur	Urban Agglomeration	680	1 433	2 116	4.7	2.8	0.1	0.3
India	Rajkot	Urban Agglomeration	974	1 647	2 322	3.3	2.5	0.1	0.4
India	Ranchi	Urban Agglomeration	844	1 293	1 793	2.7	2.3	0.1	0.3
India	Salem	Urban Agglomeration	736	1 022	1 394	2.1	2.2	0.1	0.2
India	Srinagar	Urban Agglomeration	954	1 464	2 030	2.7	2.3	0.1	0.3
India	Surat	Urban Agglomeration	2 699	5 902	8 616	4.9	2.7	0.5	1.4
India	Thiruvananthapuram	Urban Agglomeration	1 153	2 029	2 878	3.5	2.5	0.2	0.5
India	Thrissur	Urban Agglomeration	1 050	2 443	3 651	5.3	2.9	0.2	0.6
India	Tiruchirappalli	Urban Agglomeration	837	1 125	1 525	1.8	2.2	0.1	0.3
India	Tiruppur	Urban Agglomeration	523	1 295	1 974	5.7	3.0	0.1	0.3
India	Vadodara	Urban Agglomeration	1 465	2 011	2 716	2.0	2.1	0.2	0.5
India	Varanasi (Benares)	Urban Agglomeration	1 199	1 566	2 102	1.7	2.1	0.1	0.4
India	Vijayawada	Urban Agglomeration	999	1 822	2 614	3.8	2.6	0.1	0.4
India	Visakhapatnam	Urban Agglomeration	1 309	1 982	2 733	2.6	2.3	0.2	0.5
Iran (Islamic Republic of)	Ahvaz	City Proper	868	1 245	1 575	2.3	1.7	1.5	2.1
Iran (Islamic Republic of)	Esfahan	City Proper	1 382	1 915	2 364	2.0	1.5	2.4	3.2
Iran (Islamic Republic of)	Karaj	City Proper	1 087	1 861	2 386	3.4	1.8	2.3	3.1
Iran (Islamic Republic of)	Mashhad	City Proper	2 073	3 088	3 863	2.5	1.6	3.8	5.2
Iran (Islamic Republic of)	Qom	City Proper	843	1 234	1 568	2.4	1.7	1.5	2.1
Iran (Islamic Republic of)	Shiraz	City Proper	1 115	1 716	2 233	2.7	1.9	2.1	2.9
Iran (Islamic Republic of)	Tabriz	City Proper	1 264	1 594	1 943	1.4	1.4	2.0	2.7
Iran (Islamic Republic of)	Tehran	City Proper	7 128	8 516	9 990	1.1	1.1	10.6	14.3
Nepal	Kathmandu	City Proper	644	1 224	1 855	4.0	3.0	4.3	22.4
Pakistan	Faisalabad	Urban Agglomeration	2 142	3 677	5 419	3.4	2.8	1.9	4.9
Pakistan	Gujranwala	Urban Agglomeration	1 226	2 193	3 274	3.6	2.9	1.1	2.9
Pakistan	Hyderabad	Urban Agglomeration	1 221	1 812	2 613	2.5	2.6	0.9	2.4
Pakistan	Islamabad	City Proper	597	1 433	2 275	5.5	3.3	0.7	1.9
Pakistan	Karachi	Urban Agglomeration	10 032	17 121	24 838	3.3	2.7	9.0	22.8
Pakistan	Lahore	Urban Agglomeration	5 452	8 990	13 033	3.1	2.7	4.7	12.0
Pakistan	Multan	Urban Agglomeration	1 263	1 969	2 866	2.8	2.7	1.0	2.6

Country or area	City	Statistical Concepts	City population (thousands)			Average annual rate of change (percentage)		City population as a proportion of the country or area's total or urban population in 2016 (percentage)	
			2000	2016	2030	2000-2016	2016-2030	Total population	Urban population
Pakistan	Peshawar	Urban Agglomeration	1 066	1 787	2 640	3.2	2.8	0.9	2.4
Pakistan	Quetta	Urban Agglomeration	615	1 148	1 740	3.9	3.0	0.6	1.5
Pakistan	Rawalpindi	Urban Agglomeration	1 521	2 582	3 809	3.3	2.8	1.4	3.4
South-Eastern Asia									
Cambodia	Phnum Pénh (Phnom Penh)	Urban Agglomeration	1 149	1 779	2 584	2.7	2.7	11.2	53.3
Indonesia	Bandung	City Proper	2 138	2 578	3 433	1.2	2.0	1.0	1.8
Indonesia	Batam	City Proper	415	1 498	2 486	8.0	3.6	0.6	1.1
Indonesia	Bogor	City Proper	751	1 102	1 541	2.4	2.4	0.4	0.8
Indonesia	Denpasar	City Proper	409	1 177	1 870	6.6	3.3	0.5	0.8
Indonesia	Jakarta	City Proper	8 390	10 483	13 812	1.4	2.0	4.1	7.4
Indonesia	Makassar (Ujung Pandang)	City Proper	1 077	1 522	2 104	2.2	2.3	0.6	1.1
Indonesia	Medan	City Proper	1 912	2 230	2 955	1.0	2.0	0.9	1.6
Indonesia	Palembang	City Proper	1 459	1 460	1 888	0.0	1.8	0.6	1.0
Indonesia	Pekan Baru	City Proper	588	1 168	1 731	4.3	2.8	0.5	0.8
Indonesia	Semarang	City Proper	1 427	1 648	2 188	0.9	2.0	0.6	1.2
Indonesia	Surabaya	City Proper	2 611	2 878	3 760	0.6	1.9	1.1	2.0
Lao People's Dem. Republic	Vientiane	Urban Agglomeration	442	1 050	1 782	5.4	3.8	14.7	37.0
Malaysia	Kuala Lumpur	Metropolitan area	4 176	7 047	9 423	3.3	2.1	22.7	30.1
Myanmar	Mandalay	Urban Agglomeration	810	1 196	1 654	2.4	2.3	2.2	6.3
Myanmar	Nay Pyi Taw	City Proper	—	1 045	1 398	..	2.1	1.9	5.5
Myanmar	Yangon	Urban Agglomeration	3 553	4 904	6 578	2.0	2.1	9.0	25.9
Philippines	Davao City	City Proper	1 152	1 662	2 216	2.3	2.1	1.6	3.6
Philippines	Manila	Urban Agglomeration	9 962	13 131	16 756	1.7	1.7	12.7	28.6
Singapore	Singapore	Urban Agglomeration	3 918	5 717	6 578	2.4	1.0	100.0	100.0
Thailand	Krung Thep (Bangkok)	Metropolitan area	6 360	9 444	11 528	2.5	1.4	14.0	27.1
Thailand	Samut Prakan	City Proper	389	1 980	3 139	10.2	3.3	2.9	5.7
Viet Nam	Can Tho	City Proper	439	1 242	1 902	6.5	3.0	1.3	3.9
Viet Nam	Hà Nôi	Urban Agglomeration	1 660	3 790	5 498	5.2	2.7	4.0	11.8
Viet Nam	Hai Phòng	Urban Agglomeration	599	1 110	1 569	3.9	2.5	1.2	3.4
Viet Nam	Thành Pho Ho Chí Minh (Ho Chi Minh City)	Urban Agglomeration	4 389	7 498	10 200	3.3	2.2	8.0	23.3
Western Asia									

Country or area	City	Statistical Concepts	City population (thousands)			Average annual rate of change (percentage)		City population as a proportion of the country or area's total or urban population in 2016 (percentage)	
			2000	2016	2030	2000-2016	2016-2030	Total population	Urban population
Armenia	Yerevan	City Proper	1 111	1 040	1 057	-0.4	0.1	34.8	55.6
Azerbaijan	Baku	Urban Agglomeration	1 806	2 429	2 971	1.9	1.4	25.0	45.6
Georgia	Tbilisi	City Proper	1 100	1 145	1 119	0.3	-0.2	26.7	49.7
Iraq	Al-Basrah (Basra)	City Proper	759	1 041	1 491	2.0	2.6	2.8	4.1
Iraq	Al-Mawsil (Mosul)	Urban Agglomeration	1 056	1 749	2 586	3.2	2.8	4.8	6.8
Iraq	Baghdad	Metropolitan area	5 200	6 811	9 710	1.7	2.5	18.5	26.6
Iraq	Irbil (Erbil)	City Proper	757	1 200	1 766	2.9	2.8	3.3	4.7
Iraq	Sulaimaniya	Urban Agglomeration	580	1 041	1 571	3.7	2.9	2.8	4.1
Israel	Hefa (Haifa)	Metropolitan area	905	1 105	1 314	1.2	1.2	13.8	14.9
Israel	Tel Aviv-Yafo (Tel Aviv-Jaffa)	Metropolitan area	2 739	3 661	4 382	1.8	1.3	45.6	49.5
Jordan	Amman	City Proper	1 017	1 159	1 355	0.8	1.1	14.8	17.7
Kuwait	Al Kuwayt (Kuwait City)	City Proper	1 300	2 874	3 915	5.0	2.2	78.1	79.4
Lebanon	Bayrut (Beirut)	Urban Agglomeration	1 487	2 263	2 437	2.6	0.5	44.6	50.7
Saudi Arabia	Ad-Dammam	City Proper	639	1 085	1 321	3.3	1.4	3.6	4.3
Saudi Arabia	Al-Madinah (Medina)	City Proper	795	1 303	1 570	3.1	1.3	4.3	5.1
Saudi Arabia	Ar-Riyadh (Riyadh)	City Proper	3 567	6 540	7 940	3.8	1.4	21.5	25.8
Saudi Arabia	Jiddah	City Proper	2 509	4 161	4 988	3.2	1.3	13.7	16.4
Saudi Arabia	Makkah (Mecca)	City Proper	1 168	1 799	2 146	2.7	1.3	5.9	7.1
Syrian Arab Republic	Dimashq (Damascus)	Urban Agglomeration	2 017	2 586	3 451	1.6	2.1	11.4	19.6
Syrian Arab Republic	Halab (Aleppo)	Urban Agglomeration	2 204	3 641	5 087	3.1	2.4	16.0	27.5
Syrian Arab Republic	Hamah	Urban Agglomeration	495	1 297	2 003	6.0	3.1	5.7	9.8
Syrian Arab Republic	Hims (Homs)	Urban Agglomeration	856	1 695	2 471	4.3	2.7	7.4	12.8
Turkey	Adana	Urban Agglomeration	1 123	1 879	2 351	3.2	1.6	2.4	3.3
Turkey	Ankara	Urban Agglomeration	3 179	4 852	5 875	2.6	1.4	6.3	8.5
Turkey	Antalya	City Proper	595	1 100	1 381	3.8	1.6	1.4	1.9
Turkey	Bursa	Urban Agglomeration	1 180	1 974	2 465	3.2	1.6	2.5	3.4
Turkey	Gaziantep	Urban Agglomeration	844	1 567	1 955	3.9	1.6	2.0	2.7
Turkey	Istanbul	Urban Agglomeration	8 744	14 365	16 694	3.1	1.1	18.5	25.1
Turkey	Izmir	Urban Agglomeration	2 216	3 090	3 701	2.1	1.3	4.0	5.4
Turkey	Konya	Urban Agglomeration	734	1 226	1 542	3.2	1.6	1.6	2.1
United Arab Emirates	Abu Zaby (Abu Dhabi)	City Proper	505	1 179	1 608	5.3	2.2	12.1	14.1

Country or area	City	Statistical Concepts	City population (thousands)			Average annual rate of change (percentage)		City population as a proportion of the country or area's total or urban population in 2016 (percentage)	
			2000	2016	2030	2000-2016	2016-2030	Total population	Urban population
United Arab Emirates	Dubayy (Dubai)	Urban Agglomeration	907	2 504	3 471	6.3	2.3	25.7	29.9
United Arab Emirates	Sharjah	City Proper	445	1 332	1 890	6.9	2.5	13.7	15.9
Yemen	Sana'a'	Urban Agglomeration	1 347	3 094	5 071	5.2	3.5	11.9	33.7
Europe									
Eastern Europe									
Belarus	Minsk	Urban Agglomeration	1 700	1 925	1 942	0.8	0.1	20.9	27.1
Bulgaria	Sofia	Urban Agglomeration	1 128	1 230	1 230	0.5	0.0	17.4	23.5
Czech Republic	Praha (Prague)	City Proper	1 172	1 324	1 437	0.8	0.6	12.2	16.8
Hungary	Budapest	City Proper	1 787	1 712	1 811	-0.3	0.4	17.3	24.2
Poland	Warszawa (Warsaw)	City Proper	1 666	1 727	1 791	0.2	0.3	4.5	7.5
Romania	Bucuresti (Bucharest)	City Proper	1 949	1 865	1 939	-0.3	0.3	8.7	15.8
Russian Federation	Chelyabinsk	City Proper	1 082	1 160	1 187	0.4	0.2	0.8	1.1
Russian Federation	Kazan	City Proper	1 096	1 163	1 185	0.4	0.1	0.8	1.1
Russian Federation	Krasnoyarsk	City Proper	911	1 013	1 047	0.7	0.2	0.7	1.0
Russian Federation	Moskva (Moscow)	City Proper	10 005	12 260	12 200	1.3	0.0	8.7	11.7
Russian Federation	Nizhniy Novgorod	City Proper	1 331	1 200	1 060	-0.6	-0.9	0.8	1.1
Russian Federation	Novosibirsk	City Proper	1 426	1 498	1 517	0.3	0.1	1.1	1.4
Russian Federation	Omsk	City Proper	1 136	1 161	1 175	0.1	0.1	0.8	1.1
Russian Federation	Rostov-na-Donu (Rostov-on-Don)	City Proper	1 061	1 095	1 110	0.2	0.1	0.8	1.0
Russian Federation	Samara	City Proper	1 173	1 162	1 171	-0.1	0.1	0.8	1.1
Russian Federation	Sankt Peterburg (Saint Petersburg)	City Proper	4 719	5 001	4 955	0.4	-0.1	3.5	4.8
Russian Federation	Ufa	City Proper	1 049	1 069	1 085	0.1	0.1	0.8	1.0
Russian Federation	Volgograd	City Proper	1 010	1 020	1 032	0.1	0.1	0.7	1.0
Russian Federation	Yekaterinburg	City Proper	1 303	1 381	1 407	0.4	0.1	1.0	1.3
Ukraine	Kharkiv	City Proper	1 484	1 438	1 393	-0.2	-0.2	3.2	4.6
Ukraine	Kyiv (Kiev)	City Proper	2 606	2 966	3 038	0.8	0.2	6.7	9.6
Ukraine	Odesa	City Proper	1 037	1 011	1 031	-0.2	0.1	2.3	3.3
Northern Europe									
Denmark	København (Copenhagen)	Urban Agglomeration	1 077	1 281	1 455	1.1	0.9	22.5	25.7
Finland	Helsinki	Urban Agglomeration	1 019	1 190	1 293	1.0	0.6	21.7	25.7

Country or area	City	Statistical Concepts	City population (thousands)			Average annual rate of change (percentage)		City population as a proportion of the country or area's total or urban population in 2016 (percentage)	
			2000	2016	2030	2000-2016	2016-2030	Total population	Urban population
Ireland	Dublin	Urban Agglomeration	989	1 185	1 467	1.1	1.5	24.8	39.0
Norway	Oslo	Urban Agglomeration	774	1 002	1 186	1.6	1.2	19.3	23.9
Sweden	Stockholm	Urban Agglomeration	1 206	1 507	1 757	1.4	1.1	15.4	18.0
United Kingdom	Birmingham (West Midlands)	Urban Agglomeration	2 280	2 533	2 808	0.7	0.7	3.9	4.8
United Kingdom	Glasgow	Urban Agglomeration	1 180	1 227	1 360	0.2	0.7	1.9	2.3
United Kingdom	London	Urban Agglomeration	8 613	10 434	11 467	1.2	0.7	16.3	19.6
United Kingdom	Manchester	Urban Agglomeration	2 345	2 668	2 968	0.8	0.8	4.2	5.0
United Kingdom	West Yorkshire	Urban Agglomeration	1 495	1 944	2 235	1.6	1.0	3.0	3.7
Southern Europe									
Greece	Athínai (Athens)	Urban Agglomeration	3 179	3 046	3 169	-0.3	0.3	27.4	35.0
Italy	Milano (Milan)	Metropolitan area	2 985	3 104	3 162	0.2	0.1	5.1	7.3
Italy	Napoli (Naples)	Metropolitan area	2 232	2 198	2 226	-0.1	0.1	3.6	5.2
Italy	Roma (Rome)	Metropolitan area	3 385	3 738	3 842	0.6	0.2	6.1	8.8
Italy	Torino (Turin)	Metropolitan area	1 694	1 769	1 825	0.3	0.2	2.9	4.2
Portugal	Lisboa (Lisbon)	Metropolitan area	2 672	2 902	3 192	0.5	0.7	27.4	42.7
Portugal	Porto	Metropolitan area	1 254	1 304	1 443	0.2	0.7	12.3	19.2
Serbia	Beograd (Belgrade)	Urban Agglomeration	1 122	1 183	1 196	0.3	0.1	12.6	22.7
Spain	Barcelona	City Proper	4 355	5 309	5 685	1.2	0.5	11.2	14.1
Spain	Madrid	City Proper	5 014	6 264	6 707	1.4	0.5	13.2	16.6
Western Europe									
Austria	Wien (Vienna)	City Proper	1 549	1 763	1 959	0.8	0.8	20.5	31.1
Belgium	Bruxelles-Brussel	Metropolitan area	1 792	2 061	2 203	0.9	0.5	18.4	18.8
France	Lille	Urban Agglomeration	1 002	1 030	1 142	0.2	0.7	1.6	2.0
France	Lyon	Urban Agglomeration	1 443	1 622	1 814	0.7	0.8	2.5	3.1
France	Marseille-Aix-en-Provence	Urban Agglomeration	1 474	1 616	1 798	0.6	0.8	2.5	3.1
France	Paris	Urban Agglomeration	9 737	10 925	11 803	0.7	0.6	16.7	21.0
Germany	Berlin	City Proper	3 384	3 578	3 658	0.3	0.2	4.3	5.7
Germany	Hamburg	City Proper	1 710	1 839	1 906	0.5	0.3	2.2	3.0
Germany	Köln (Cologne)	City Proper	963	1 042	1 095	0.5	0.4	1.3	1.7
Germany	München (Munich)	City Proper	1 202	1 454	1 548	1.2	0.4	1.8	2.3
Netherlands	Amsterdam	Urban Agglomeration	1 005	1 099	1 213	0.6	0.7	6.5	7.1

Country or area	City	Statistical Concepts	City population (thousands)			Average annual rate of change (percentage)		City population as a proportion of the country or area's total or urban population in 2016 (percentage)	
			2000	2016	2030	2000-2016	2016-2030	Total population	Urban population
Switzerland	Zürich (Zurich)	Urban Agglomeration	1 078	1 259	1 494	1.0	1.2	15.1	20.5
Latin America and the Caribbean									
Caribbean									
Cuba	La Habana (Havana)	Metropolitan area	2 186	2 129	2 104	-0.2	-0.1	18.9	24.5
Dominican Republic	Santo Domingo	Urban Agglomeration	1 997	3 020	3 888	2.6	1.8	28.0	35.1
Haiti	Port-au-Prince	Urban Agglomeration	1 693	2 507	3 525	2.5	2.4	23.3	39.0
Puerto Rico	San Juan	Metropolitan area	2 508	2 460	2 468	-0.1	0.0	66.9	71.5
Central America									
Costa Rica	San José	Urban Agglomeration	1 032	1 183	1 501	0.9	1.7	23.4	30.1
El Salvador	San Salvador	Metropolitan area	1 062	1 102	1 254	0.2	0.9	17.0	25.4
Guatemala	Ciudad de Guatemala (Guatemala City)	Urban Agglomeration	1 973	2 994	4 650	2.6	3.1	18.0	34.6
Honduras	Tegucigalpa	City Proper	752	1 146	1 613	2.6	2.4	13.3	24.1
Mexico	Aguascalientes	Metropolitan area	763	1 050	1 287	2.0	1.5	0.8	1.0
Mexico	Ciudad de México (Mexico City)	Metropolitan area	18 457	21 157	23 865	0.9	0.9	16.7	21.0
Mexico	Ciudad Juárez	Metropolitan area	1 223	1 401	1 650	0.9	1.2	1.1	1.4
Mexico	Cuernavaca	Metropolitan area	803	1 006	1 211	1.4	1.3	0.8	1.0
Mexico	Guadalajara	Metropolitan area	3 724	4 920	5 837	1.7	1.2	3.9	4.9
Mexico	León de los Aldamas	Metropolitan area	1 280	1 845	2 260	2.3	1.5	1.5	1.8
Mexico	Mérida	Metropolitan area	810	1 086	1 323	1.8	1.4	0.9	1.1
Mexico	Mexicali	Metropolitan area	770	1 053	1 288	2.0	1.4	0.8	1.0
Mexico	Monterrey	Metropolitan area	3 405	4 589	5 471	1.9	1.3	3.6	4.6
Mexico	Puebla	Metropolitan area	2 285	3 032	3 628	1.8	1.3	2.4	3.0
Mexico	Querétaro	Metropolitan area	825	1 300	1 630	2.8	1.6	1.0	1.3
Mexico	San Luis Potosí	Metropolitan area	857	1 168	1 426	1.9	1.4	0.9	1.2
Mexico	Tijuana	Metropolitan area	1 365	2 032	2 502	2.5	1.5	1.6	2.0
Mexico	Toluca de Lerdo	Metropolitan area	1 553	2 207	2 690	2.2	1.4	1.7	2.2
Mexico	Torreón	Metropolitan area	1 014	1 354	1 643	1.8	1.4	1.1	1.3
Panama	Ciudad de Panamá (Panama City)	Urban Agglomeration	1 216	1 708	2 221	2.1	1.9	42.2	63.1
South America									
Argentina	Buenos Aires	Urban Agglomeration	12 407	15 334	16 956	1.3	0.7	36.1	39.3
Argentina	Córdoba	Urban Agglomeration	1 348	1 519	1 718	0.7	0.9	3.6	3.9

Country or area	City	Statistical Concepts	City population (thousands)			Average annual rate of change (percentage)		City population as a proportion of the country or area's total or urban population in 2016 (percentage)	
			2000	2016	2030	2000-2016	2016-2030	Total population	Urban population
Argentina	Mendoza	Urban Agglomeration	838	1 020	1 181	1.2	1.0	2.4	2.6
Argentina	Rosario	Urban Agglomeration	1 152	1 395	1 607	1.2	1.0	3.3	3.6
Bolivia (Plurinational State of)	Cochabamba	Metropolitan area	721	1 273	1 703	3.6	2.1	11.4	16.5
Bolivia (Plurinational State of)	La Paz	Urban Agglomeration	1 390	1 834	2 308	1.7	1.6	16.4	23.8
Bolivia (Plurinational State of)	Santa Cruz	City Proper	1 054	2 181	2 989	4.5	2.3	19.5	28.3
Brazil	Baixada Santista	Urban Agglomeration	1 309	1 551	1 769	1.1	0.9	0.8	0.9
Brazil	Belém	Metropolitan area	1 724	2 209	2 540	1.5	1.0	1.1	1.3
Brazil	Belo Horizonte	Metropolitan area	4 807	5 766	6 439	1.1	0.8	2.8	3.3
Brazil	Brasília	Metropolitan area	2 932	4 235	4 929	2.3	1.1	2.1	2.4
Brazil	Campinas	Metropolitan area	2 332	3 091	3 560	1.8	1.0	1.5	1.8
Brazil	Curitiba	Metropolitan area	2 494	3 537	4 116	2.2	1.1	1.7	2.0
Brazil	Florianópolis	Metropolitan area	734	1 212	1 481	3.1	1.4	0.6	0.7
Brazil	Fortaleza	Metropolitan area	2 875	3 944	4 551	2.0	1.0	1.9	2.2
Brazil	Goiânia	Metropolitan area	1 635	2 327	2 730	2.2	1.1	1.1	1.3
Brazil	Grande São Luís	Urban Agglomeration	1 064	1 460	1 714	2.0	1.1	0.7	0.8
Brazil	Grande Vitória	Urban Agglomeration	1 314	1 655	1 908	1.4	1.0	0.8	0.9
Brazil	João Pessoa	Urban Agglomeration	842	1 109	1 298	1.7	1.1	0.5	0.6
Brazil	Joinville	Metropolitan area	923	1 237	1 450	1.8	1.1	0.6	0.7
Brazil	Maceió	Metropolitan area	952	1 286	1 508	1.9	1.1	0.6	0.7
Brazil	Manaus	City Proper	1 392	2 069	2 454	2.5	1.2	1.0	1.2
Brazil	Natal	Urban Agglomeration	860	1 186	1 399	2.0	1.2	0.6	0.7
Brazil	Pôrto Alegre	Metropolitan area	3 216	3 621	4 028	0.7	0.8	1.8	2.1
Brazil	Recife	Metropolitan area	3 205	3 767	4 222	1.0	0.8	1.8	2.1
Brazil	Rio de Janeiro	Metropolitan area	11 307	12 981	14 174	0.9	0.6	6.3	7.4
Brazil	Salvador	Metropolitan area	2 891	3 623	4 115	1.4	0.9	1.8	2.1
Brazil	São Paulo	Metropolitan area	17 014	21 297	23 444	1.4	0.7	10.4	12.1
Chile	Santiago	Urban Agglomeration	5 658	6 544	7 122	0.9	0.6	36.2	40.4
Colombia	Barranquilla	Metropolitan area	1 531	2 009	2 370	1.7	1.2	4.0	5.2
Colombia	Bogotá	Urban Agglomeration	6 356	9 968	11 966	2.8	1.3	19.9	25.9
Colombia	Bucaramanga	Urban Agglomeration	855	1 235	1 507	2.3	1.4	2.5	3.2
Colombia	Cali	Urban Agglomeration	1 950	2 682	3 203	2.0	1.3	5.4	7.0

Country or area	City	Statistical Concepts	City population (thousands)			Average annual rate of change (percentage)		City population as a proportion of the country or area's total or urban population in 2016 (percentage)	
			2000	2016	2030	2000-2016	2016-2030	Total population	Urban population
Colombia	Cartagena	City Proper	737	1 113	1 372	2.6	1.5	2.2	2.9
Colombia	Medellín	Metropolitan area	2 724	3 972	4 747	2.4	1.3	7.9	10.3
Ecuador	Guayaquil	Urban Agglomeration	2 077	2 756	3 493	1.8	1.7	16.7	26.2
Ecuador	Quito	City Proper	1 357	1 754	2 228	1.6	1.7	10.7	16.6
Paraguay	Asunción	Urban Agglomeration	1 499	2 406	3 135	3.0	1.9	33.7	56.2
Peru	Lima	Metropolitan area	7 294	10 072	12 221	2.0	1.4	31.9	40.4
Uruguay	Montevideo	Metropolitan area	1 600	1 716	1 860	0.4	0.6	49.9	52.2
Venezuela (Bolivarian Republic of)	Barquisimeto	Metropolitan area	946	1 044	1 228	0.6	1.2	3.3	3.7
Venezuela (Bolivarian Republic of)	Caracas	Metropolitan area	2 864	2 923	3 347	0.1	1.0	9.2	10.3
Venezuela (Bolivarian Republic of)	Maracaibo	Metropolitan area	1 724	2 229	2 677	1.6	1.3	7.0	7.9
Venezuela (Bolivarian Republic of)	Maracay	Metropolitan area	898	1 186	1 450	1.7	1.4	3.7	4.2
Venezuela (Bolivarian Republic of)	Valencia	Metropolitan area	1 392	1 757	2 107	1.5	1.3	5.5	6.2
Northern America									
Canada	Calgary	Metropolitan area	927	1 365	1 630	2.4	1.3	3.8	4.6
Canada	Edmonton	Metropolitan area	924	1 298	1 547	2.1	1.3	3.6	4.4
Canada	Montréal	Metropolitan area	3 429	4 014	4 517	1.0	0.8	11.1	13.5
Canada	Ottawa-Gatineau	Metropolitan area	1 055	1 346	1 577	1.5	1.1	3.7	4.5
Canada	Toronto	Metropolitan area	4 607	6 083	6 957	1.7	1.0	16.8	20.5
Canada	Vancouver	Metropolitan area	1 959	2 523	2 930	1.6	1.1	7.0	8.5
United States of America	Atlanta	Urban Agglomeration	3 522	5 249	6 140	2.5	1.1	1.6	2.0
United States of America	Austin	Urban Agglomeration	911	1 744	2 182	4.1	1.6	0.5	0.7
United States of America	Baltimore	Urban Agglomeration	2 079	2 275	2 543	0.6	0.8	0.7	0.8
United States of America	Boston	Urban Agglomeration	4 036	4 255	4 671	0.3	0.7	1.3	1.6
United States of America	Charlotte	Urban Agglomeration	768	1 685	2 166	4.9	1.8	0.5	0.6
United States of America	Chicago	Urban Agglomeration	8 315	8 755	9 493	0.3	0.6	2.7	3.3
United States of America	Cincinnati	Urban Agglomeration	1 506	1 696	1 916	0.7	0.9	0.5	0.6
United States of America	Cleveland	Urban Agglomeration	1 786	1 769	1 948	-0.1	0.7	0.5	0.7
United States of America	Columbus, Ohio	Urban Agglomeration	1 139	1 528	1 788	1.8	1.1	0.5	0.6
United States of America	Dallas-Fort Worth	Urban Agglomeration	4 168	5 799	6 683	2.1	1.0	1.8	2.2
United States of America	Denver-Aurora	Urban Agglomeration	1 994	2 636	3 048	1.7	1.0	0.8	1.0
United States of America	Detroit	Urban Agglomeration	3 899	3 618	3 886	-0.5	0.5	1.1	1.4

Country or area	City	Statistical Concepts	City population (thousands)			Average annual rate of change (percentage)		City population as a proportion of the country or area's total or urban population in 2016 (percentage)	
			2000	2016	2030	2000-2016	2016-2030	Total population	Urban population
United States of America	Houston	Urban Agglomeration	3 847	5 757	6 729	2.5	1.1	1.8	2.1
United States of America	Indianapolis	Urban Agglomeration	1 225	1 672	1 961	1.9	1.1	0.5	0.6
United States of America	Jacksonville, Florida	Urban Agglomeration	886	1 190	1 398	1.8	1.2	0.4	0.4
United States of America	Kansas City	Urban Agglomeration	1 365	1 617	1 846	1.1	0.9	0.5	0.6
United States of America	Las Vegas	Urban Agglomeration	1 326	2 340	2 867	3.5	1.5	0.7	0.9
United States of America	Los Angeles-Long Beach-Santa Ana	Urban Agglomeration	11 798	12 317	13 257	0.3	0.5	3.8	4.6
United States of America	Louisville	Urban Agglomeration	866	1 041	1 201	1.2	1.0	0.3	0.4
United States of America	Memphis	Urban Agglomeration	974	1 113	1 269	0.8	0.9	0.3	0.4
United States of America	Miami	Urban Agglomeration	4 933	5 864	6 554	1.1	0.8	1.8	2.2
United States of America	Milwaukee	Urban Agglomeration	1 311	1 413	1 588	0.5	0.8	0.4	0.5
United States of America	Minneapolis-St. Paul	Urban Agglomeration	2 395	2 812	3 175	1.0	0.9	0.9	1.0
United States of America	Nashville-Davidson	Urban Agglomeration	755	1 129	1 357	2.5	1.3	0.3	0.4
United States of America	New York-Newark	Urban Agglomeration	17 813	18 604	19 885	0.3	0.5	5.7	6.9
United States of America	Orlando	Urban Agglomeration	1 165	1 769	2 115	2.6	1.3	0.5	0.7
United States of America	Philadelphia	Urban Agglomeration	5 156	5 602	6 158	0.5	0.7	1.7	2.1
United States of America	Phoenix-Mesa	Urban Agglomeration	2 923	4 135	4 808	2.2	1.1	1.3	1.5
United States of America	Pittsburgh	Urban Agglomeration	1 753	1 715	1 884	-0.1	0.7	0.5	0.6
United States of America	Portland	Urban Agglomeration	1 589	2 025	2 336	1.5	1.0	0.6	0.8
United States of America	Providence	Urban Agglomeration	1 175	1 196	1 333	0.1	0.8	0.4	0.4
United States of America	Raleigh	Urban Agglomeration	548	1 188	1 533	4.8	1.8	0.4	0.4
United States of America	Richmond	Urban Agglomeration	822	1 042	1 214	1.5	1.1	0.3	0.4
United States of America	Riverside-San Bernardino	Urban Agglomeration	1 516	2 239	2 653	2.4	1.2	0.7	0.8
United States of America	Sacramento	Urban Agglomeration	1 401	1 953	2 294	2.1	1.1	0.6	0.7
United States of America	Salt Lake City	Urban Agglomeration	891	1 108	1 284	1.4	1.1	0.3	0.4
United States of America	San Antonio	Urban Agglomeration	1 337	2 077	2 488	2.8	1.3	0.6	0.8
United States of America	San Diego	Urban Agglomeration	2 681	3 129	3 522	1.0	0.8	1.0	1.2
United States of America	San Francisco-Oakland	Urban Agglomeration	3 230	3 299	3 615	0.1	0.7	1.0	1.2
United States of America	San Jose	Urban Agglomeration	1 541	1 739	1 964	0.8	0.9	0.5	0.6
United States of America	Seattle	Urban Agglomeration	2 720	3 278	3 709	1.2	0.9	1.0	1.2
United States of America	St. Louis	Urban Agglomeration	2 079	2 187	2 427	0.3	0.7	0.7	0.8
United States of America	Tampa-St. Petersburg	Urban Agglomeration	2 071	2 694	3 105	1.6	1.0	0.8	1.0

Country or area	City	Statistical Concepts	City population (thousands)			Average annual rate of change (percentage)		City population as a proportion of the country or area's total or urban population in 2016 (percentage)	
			2000	2016	2030	2000-2016	2016-2030	Total population	Urban population
United States of America	Virginia Beach	Urban Agglomeration	1 396	1 462	1 632	0.3	0.8	0.4	0.5
United States of America	Washington, D.C.	Urban Agglomeration	3 949	5 013	5 690	1.5	0.9	1.5	1.9
Oceania									
Australia/New Zealand									
Australia	Adelaide	Metropolitan area	1 142	1 265	1 505	0.6	1.2	5.2	5.8
Australia	Brisbane	Metropolitan area	1 666	2 238	2 721	1.8	1.4	9.2	10.3
Australia	Melbourne	Metropolitan area	3 461	4 258	5 071	1.3	1.2	17.6	19.6
Australia	Perth	Metropolitan area	1 432	1 896	2 329	1.8	1.5	7.8	8.7
Australia	Sydney	Metropolitan area	4 052	4 540	5 301	0.7	1.1	18.7	20.9
New Zealand	Auckland	Urban Agglomeration	1 063	1 360	1 574	1.5	1.0	29.3	34.0

Data source: United Nations, Department of Economic and Social Affairs, Population Division (2014). *World Urbanization Prospects: The 2014 Revision.*

Notes

This annex table includes only cities with 1 million inhabitants or more on 1 July 2016.

An em dash (—) indicates that the value is zero (magnitude zero).

Two dots (..) indicate that the item is not applicable.

A minus sign (-) before a figure indicates a decrease

A full stop is used to indicate decimals.

Use of en dash (–) between years, for example 2000-2016, signifies the full period involved, from 1 July of the beginning year to 1 July of the end year.

www.ingramcontent.com/pod-product-compliance
Lightning Source LLC
Chambersburg PA
CBHW060911270326

41933CB00004B/210

9 789211 515497